It sprang from its Awareness

How to become a
Cosmopsychist
with Kabbalah and Quanta.

Axel B.C. Krauss

Copyright © 2024 Axel B.C. Krauss
All rights reserved.
ISBN: 9798333908308

"What many people think is a conflict between religion and science is actually something else. It is a conflict between religion and materialism. Materialism regards itself as scientific and is often labelled »scientific materialism« even by its opponents, but it has no legitimate claim to be part of science. Rather, it is a school of philosophy characterised by the belief that nothing exists apart from matter, or, as Democritus put it, »atoms and nothingness«."

Stephen Barr, Professor of Physics and Astronomy
at the University of Delaware and author of the book
"Modern Physics and Ancient Faith"

„Our entire universe is but a projection of mind-idea into a three-dimensional universe. Here, thought waves become perceptible in temporal succession from the universal zero point of rest, and this if followed by a cancellation of this projection by pulling it back into the universal zero point of rest."

Walter Russell, „Eine neue Vorstellung vom Universum",
(A new concept of the universe), Genius-Verlag 2019,
p. 112

*„… as the seed is the folded tree,
so the world is the unfolded god."*

Manly P. Hall, „The Secret Teachings of All Ages",
H.S. Crocker Company, 1928, p. 169

ACKNOWLEDGMENTS

Infinite thanks and deepest respect to all the thinkers and scientists who have expanded our understanding of the cosmos and our existence through their tireless work. Above all, I am indebted to Sir Roger Penrose, whose book "Shadows of the Mind" provided me with many ideas and great inspiration. And Walter Russell's book "A New Concept of the Universe" opened up perspectives for me without which I would probably not have found the central ideas of this short, humble essay of mine. I am only a modest philosopher, but I hope that I will perhaps provide one or two interesting suggestions to the readers. If so, I'd be satisfied.

I
"Spirit" and "Matter": A false dualism?

"Matter is not made up of matter! [...] At the end of all the splitting up of matter, something remains that is more akin to the spiritual - holistic, open, alive: Potentiality, the possibility of realization. Matter is the slag of this spirituality - decomposable, definable, determined: Reality."

Hans-Peter Dürr, physicist and essayist,
former director of the german Max Planck Institute of Physics,
„Warum es ums Ganze geht", oekom-Verlag, p. 86/87

"I regard consciousness as fundamental. I consider matter to be derived from consciousness. We cannot get beyond consciousness. Everything we talk about, everything we consider to exist, presupposes consciousness."

Max Planck, from an interview with J.W.N. Sullivan,
"The Observer", 25th January 1931, p. 17

Foreword

When I began to study the findings of modern quantum physics about five years ago, I regularly had "aha" experiences. What kind? Well, I'm actually a philosopher. Although I had a certain background in the natural sciences, I had a lot of catching up to do, especially when it came to quantum physics. Before this research, I was interested in the mystery schools and religions of antiquity for a long time. I read everything I could get my hands on about them. When I then immersed myself in quantum physics, I realised that there are quite astonishing parallels between their concepts and ideas about the nature of matter and our physical reality and the cosmogonies of, for example, Hinduism, Buddhism, Taoism and above all those of the Kabbalah.

How can that be? I asked myself whether it could all just be a coincidence or whether I had simply imagined something. That I was seeing things that weren't there and had made unjustified comparisons. Not at all. The more I compared the religious and natural and existential philosophical writings of antiquity with the statements of quantum physics, the more parallels I discovered.

This modest book is intended to show precisely these parallels. It is not a scientific treatise, but a philosophical excursion into the question of the reason and nature of existence with the aid of scientific findings. When an acquaintance asked about the length of the book and I told him that it will have just about 130 pages, he reacted with some disappointment: "What kind of book is that?"

I've often made this mistake. As a philosopher, I suspect that it is probably due to the 'mass age' in which we live: Something only has substance, something is only worthy of attention or important if it has 500 pages, if a film has grossed billions at the global box office or if some influencer on YouTube has millions of "followers". What a load of rubbish.

For me, it's the content that counts - nothing more. I have said what I wanted to say on this topic. I've read writings that were just 70 or 80 pages long, but the content was so to the point, concise and witty that I always prefer such "small books" to those that only spread boredom and thoughtlessness over four or five hundred pages.

Perhaps I have failed with this. I'll leave that judgement up to the reader. But what I definitely didn't want was an artificially bloated book just to achieve a high page count and thus drive up the price.

There are two types of quotations in the text: Those in the body text, which refer directly to the issues discussed, and those framed by dividing lines: They do not always relate directly to the text, but are intended to emphasise the intellectual "thrust"; they can be understood as "signposts" intended to underpin the overall context.

I hope that reading this book will stimulate your thoughts or perhaps even inspire you to do your own research. In this sense, this small booklet should be read as an "appetizer".

This is my first book in English. I therefore apologise if, despite my best efforts to avoid them, some grammatical and punctuation errors have crept in or if some of the wording seems a little "off".

Axel B.C. Krauss,

23.7.2024

Originally, the subtitle of this book was supposed to be: „Why belief in God and the theory of evolution are compatible. Existential philosophical reflections on a trivial question." But I quickly refrained from doing so. Because we live in an age in which the mere mention of the concept of God is almost considered disreputable; anyone who talks like that cannot possibly be serious. Who believes that the notion of a „God", evolution and – can you believe this nonsense? – quantum physics could be compatible is a certified nutjob!

Some readers would have certainly been offended by such a title. The author must have lost his mind: the question of why belief in a "God" could be compatible with the theory of evolution is "trivial". Excuse me? A question that has been causing quite clever minds to despair for a very long time and over which they regularly get into each other's hair is supposed to be trivial? So the author thinks he is smarter than all these very smart people?

No. Of course not. Why do I still claim that this is a trivial question?

Because - as should be obvious - the answer to it depends on the meaning you give to an empty term like "God". What *exactly*, what **specifically** can I imagine it to mean? What is "God"? A "creator", a "creative principle", a kind of "consciousness", a form of "intelligent energy", or what? The British theoretical physicist and astrophysicist Stephen Hawking (1942 - 2018) once answered this question as follows:

> *"You could define God as the embodiment of the laws of nature. However, that is not what most people imagine God to be. They mean a human-like being with whom you can have a personal relationship. When you consider the immense size of the universe and how insignificant and random human life is in it, this seems highly unlikely. I use the word »God« in an impersonal sense, as Einstein did for the laws of nature, so that knowing the mind of God means knowing the laws of nature. My prediction is that we will know the mind of God by the end of this century."*[1]

Later, however, Hawking expressed himself much more clearly on this question:

> *"In short, do we need a God to make the Big Bang possible? I don't want to offend anyone who is a believer, but I think science has a more convincing explanation than a divine*

---

[1] Maria Popova, „Is There a God? Stephen Hawking Gives the Definitive Answer to the Eternal Question", https://www.themarginalian.org/2019/07/17/stephen-hawking-brief-answers-to-the-big-questions/.

creator."[2]

With all due respect to an extraordinarily intelligent person like Hawking: Who says that the creation of the universe - let's just call it existence in general - must have been based on a conscious act of creation? Oh, are there also "unconscious" ones? This is getting more and more absurd! Take it easy. I ask for your patience, because I'm going to go into much more detail on this question later on.

Possibly - to only vaguely hint at it right at the beginning of this book and thus certainly invoke even more head-shaking - the emergence of existence was a **logically necessary process** from a state of non-existence. To use another formulation: did the mere probability, the possibility of existence already *necessitate* its existential concretisation or manifestation as an indispensable antithesis for the purpose of definitional precision in the form of "self-observation" or "self-measurement"?

Admittedly, that was certainly very abstract and difficult to grasp. Perhaps it would be better at this early stage to ask a very simple existential philosophical question, as children would ask: **What would be the point** of non-existence in the form of pure probabilities of existential realisation if it only ever remains in this state of "probalistic", i.e. mere probability? What would I gain from existing as a mere possibility of an author, as a "maybe-author", if I never get the opportunity to produce something that identifies me as a *concrete* author? What would be the point of the mere probability of a universe that dozes on forever without ever being existentially concretised? Is there an answer to that question? Maybe there is, but we'll come back to that at the end of the first part of this book.

> *"In fact, for the discerning, it is basically a philosophical atheism, because this supreme deity is in reality neither a personality nor a principle, but the principle of all principles, the most abstract of the most abstract, which is so universalised and unlimited in its essence that it is incomprehensible. If one compares such a deity with the popular theological idea of a personal God, the superiority of the God of philosophy becomes immediately clear. The god of*

---

[2] Ibid.

> *ancient philosophy is the deity whose sufficiency will be confirmed by modern science."*[3]

If you're a little confused or even wondering whether I'm just making fun of you: No, I'm completely serious about these questions. And of course I am by no means alone in this and certainly not the first. Questions of this kind have always preoccupied mankind. However, I believe there will never be a definitive answer - at least it is highly unlikely. For reasons that I will also go into.

> *"Don't keep saying to yourself, if you can avoid it: »But how can it be like this?«, because you will end up in a dead end from which no one has ever escaped. Nobody knows how it can be like this."*[4]

So if I now start from the assumption that "God" (if one wants to use this term) is a creative, intelligent consciousness - why should it then be easily compatible with the prevailing doctrine of gradual evolution? Are the two not necessarily mutually exclusive? Isn't that exactly what many scientists constantly claim, especially those from so-called "scientific materialism"? Moreover: Surely anyone who assumes a conscious "act of creation" as the reason for the existence of the universe belongs to the faction of "creationists" who believe that the world was created a few thousand years ago?

And if the universe, and therefore all life, is based on a conscious act of creation, doesn't that rule out "free will", because such a universe would be predeterministic, i.e. everything is predetermined according to a "divine plan"?

> *"Quantum mechanical tunneling has been observed in synaptic and enhaptic transmission [...].. The ability of the*

---

[3] Manly P. Hall, "Lectures on Ancient Philosophy", Philosophical Research Society, 1985, Jeremy T. Archer Edition, Penguin Books 2005, p. 218
[4] Richard P. Feynman, „Probability and Uncertainty: The Quantum-Mechanical View of Nature", Penguin Books, p. 111.

> observer to choose actuals from possible states is the mechanism by which agency and free will are possible."⁵

I don't believe anything like that. I am neither a creationist, nor do I believe in a personalised God, i.e. a "humanised" one. I do not believe that everything is predetermined, and I also do not believe that existence must necessarily be based on a conscious act of creation, but that "creation" can be a process of ***becoming*** self-aware. In other words: I suspect that the "driving forces" for existence may arise from the tension between consciousness and self-awareness. I can imagine that evolution could be the expression of such a process of becoming more and more self-awarene through time and therefore does not necessarily amount to predetermination, i.e. does not have to exclude "free will".

> "When we think about this question, we should remember that we are questioning destiny, not providence. There is a fundamental difference between a point of origin and a point of destiny. A point of origin is in the past and must be assumed to be a definite and unique state. A destiny point will also be a specific and unique state when it is reached - but it will not be until it is **reached**. Similar to the multipotentiality of the quantum, which can freely choose its real state from its virtual states until an interaction collapses its wave function, the cosmos will not have a definite final state until it **actually reaches** that state. **Since it is not classically mechanistic, it is indeterminate as far as the choice of its final state is concerned.**"⁶

Oh boy. Lots of puzzling formulations. How is it all supposed to fit together? And without any contradictions? Is the author of this book trying to bullshit me?

---

[5] Perry Marshall, https://www.sciencedirect.com/science/article/pii/S007961072300041X#bib3

[6] Ervin Laszlo, "Science and the Akashic Field. An Integral Theory of Everything", Inner Traditions, Vermont, 2007, p. 91.

> *"My religiosity consists in a humble admiration of the infinitely superior spirit that reveals itself in the little that we are able to recognise of reality with our weak and feeble reason."*[7]

As I said, everything depends on the concrete definition of the terms. Which is why I ask you right at the beginning not to get doggedly attached to such man-made terms. Many of these words, which are still used today as if they were self-evident and needed no further explanation, originate from ancient ideas - of "heaven" and "earth", of the "spiritual sphere" and the "material sphere", be they early religious and cosmogonic (world-creation mystical) belief systems, philosophical or - in later times - scientific. They have grown historically - and can therefore naturally also be subject to historical change. Ultimately, they are attempts by human thought to explain things for which there is still no "final", all-encompassing answer (if there can be one at all); in general, it is an attempt by humans to approach their natural environment in an analytical and descriptive way, to systematise and categorise it; one could also say that the complexity of the world is "broken down" to human levels - in the form of symbols. Human language - whether scientific or everyday or colloquial - naturally also has a symbolic character.

After all, as a human being, I can only ever make judgements within the knowledge horizon of the time in which I live. But this is not a fixed point. It is not the "last word in wisdom", it is not the "ultima ratio" - that should go without saying. To come straight to the first example: The Big Bang theory in its current form is certainly not the best or most plausible, most conclusive or most coherent "of all time".

A nonsensical formulation, by the way: "of all times". It is irrational. If you're now thinking "Well, you're probably a bean counter! That's splitting hairs!", you would have already made a cardinal mistake: simply accepting certain terms that arose from contemporary circumstances as "normal" or self-evident. I didn't choose this particular example, which is so popular these days, for nothing. Because it is an excellent illustration of what I am concerned with in this book: **the precision of conceptual thinking**, regardless of whether these are scientific or (natural or existential) philosophical concepts.

The phrase "all times" is of course absurd because it implies that one can draw conclusions about "all" times from an event in one's own time horizon. No one is capable of doing this. A simple example: "X is the most successful film of all time!".

---

[7] Albert Einstein, "Briefe. Aus dem Nachlass", published by H. Dukas and B. Hoffmann, Zurich 1981, p. 63.

But what if a film is released in the same year, or the next year if you like, that is even more successful? What do I mean by the "best comedy of all time"? And what if people laugh even louder at a comedy film that comes out in the future? Or if a future horror film scares them even more than the "scariest film of all time"? Obviously enough, these are very vague, sometimes irrational and illogical terms. And it is precisely these that should be avoided here as far as possible. Because if you use them, you don't even need to approach the questions posed in this book. Particularly in the natural sciences and humanities, the accuracy of the terminology used is of fundamental importance. It is not for nothing that many a problem in the history of philosophy has turned out to be only a supposed problem - due to linguistic inadequacies. After all, linguistic concepts are not reality - they are symbolic representations of it.

> *"Because our representation of reality is so much easier to grasp than reality itself, we tend to confuse the two and mistake our concepts and symbols for reality."*[8]

The relationship between human thought in the search for explanations of natural phenomena, be they macroscopic or microscopic in nature, and the "material reality", which for a long time was assumed to be an "objectively" fixed external world, changed in the course of the history of the natural sciences through new, sometimes revolutionary findings or ideas - especially with the advent of quantum physics. The boundary between the strict classical scientific division of the world into subjectivity and objectivity became blurred: the researcher was no longer simply an "external" or uninvolved observer scrutinising a world that was objectively opposite him, but - to put it simply - an interaction between the observer and the observed was established.

> *"Nothing is more important about the quantum principle than the fact that it destroys the idea of the world »out there«, from which the observer is separated by a 20-centimetre-thick pane of glass. Even to observe an object as tiny as an electron, he has to break the glass. He has to reach in. He has to install the measuring apparatus he has chosen. It is up to him to decide whether he wants to measure the position or the momentum. Installing the measuring equipment for one prevents and excludes the installation of the equipment for measuring the*

---

[8] Fritjof Capra, "The Tao of Physics. An Exploration of the Parallels between Modern Physics and Eastern Mysticism", Flamingo, 1982, p. 35.

> *other. In addition, the measurement changes the state of the electron. The universe will never be the same again. To describe what has happened, you have to cross out the old word »observer« and put the new word »participant« in its place. In a strange sense, the universe is a participatory universe."*[9]

In addition, other physical convictions from classical Netwon physics, which were based on the assumption of a fixed, static space, an independent, evenly running time and immovable mechanical relations, were shaken to their foundations by Einstein's general theory of relativity and quantum theory.

> *"(...) the theories of atomic and subatomic physics made the existence of elementary particles increasingly unlikely. They revealed **a fundamental relationship of matter**, showed that kinetic energy can be converted into mass, and suggested that particles **are processes rather than objects.**"*[10]

It would go too far for this book to deliver all the mathematical details, so I will limit myself to a rough outline: In the so-called "S-matrix theory", whose beginnings go back to the research of the English mathematical and theoretical physicist Paul Dirac on collisions between particles (in his 1927 paper "On the Quantum Mechanics of Collisions"), attempts are made to relate the initial and final states of a physical system undergoing a scattering process. The concept of the S-matrix was first introduced by the American theoretical physicist John Archibald Wheeler and further developed somewhat later by the German physicist Werner Heisenberg. The letter S stands for scattering, i.e. the phenomenon that can be observed in a cloud chamber where particles that are shot at each other at extremely high speeds (in a particle accelerator such as CERN) show a scattering pattern. It is therefore a theory from high-energy particle physics; it is used in quantum mechanics, scattering theory and quantum field theory (QFT). The extremely complicated mathematical details can be omitted for the discussion of this book – but the philosophical implications, which were described by Fritjof Capra in his book "The Tao of Physics" as follows,

---

[9] J.A. Wheeler, in "The Physicist's Conception of Nature", published by J. Mehra, D. Reidel, Dordrecht, Holland, 1973, p. 244

[10] Fritjof Capra, "The Tao of Physics. An Exploration of the Parallels between Modern Physics and Eastern Mysticism", Flamingo, 1982, p. 315

are decisive:

> "Such a theory of subatomic particles reflects in its most extreme form the **impossibility of separating the scientific observer from the observed phenomena**, which has already been discussed in the context of quantum theory. **It ultimately implies that the structures and phenomena we observe in nature are nothing other than creations of our measuring and categorising minds.** That this is so is one of the fundamental teachings of Eastern philosophy. Eastern mystics tell us again and again that all things and events that we perceive are creations of the mind that arise from a certain state of consciousness and dissolve again when this state is transcended."[11]

At this point, I would also like to add another extremely interesting piece of information that will become important later - namely in a discussion of the so-called panpsychist or cosmopsychist position (as part of an attempt to explain human consciousness), which I myself hold. It is based on the so-called "bootstrap" theory, which is to be seen in connection with the aforementioned S-matrix. Here, too, I will avoid overly complicated details, but only mention the basic features of this bootstrap theory: At its core is the idea that one particle represents a kind of "mirror" of all other particles and potentially contains them within itself, while all other particles in turn contain this one particle within themselves. All particles therefore potentially contain each other. Of course, that was a very simplified expression. I will therefore once again refer back to the Austrian-American physicist, systems theorist and philosopher Fritjof Capra, who, as a man of expertise, can explain the basics much better than I can. A brief explanation in advance: "Hadrons" are subatomic particles that can form protons or neutrons, for example, and are themselves made up of quarks.[12]

> "The picture of hadrons that emerges from these bootstrap models is often summarised in the provocative sentence: »Every particle consists of all other particles«. However, one should not imagine that each hadron contains all the others in a classical, static sense. Hadrons do not »contain« each other,

---

[11] Fritjof Capra, "The Tao of Physics. An Exploration of the Parallels between Modern Physics and Eastern Mysticism", Flamingo, 1982, p. 306

[12] "The use of particle accelerators made it possible to venture into the interior of protons: It was discovered that these particles consist of even smaller particles, the quarks - which we now regard as elementary particles. This was the foundation of hadron physics. It deals with the study of all particles that contain quarks and are known as hadrons.", https://www.weltderphysik.de/gebiet/teilchen/hadronen-und-kernphysik/hadronen/, my translation.

> *but »involve« each other in the dynamic and probabilistic sense of S-matrix theory, where each hadron is a potential »bound state« of all sets of particles that can interact with each other to form the hadron under consideration."*[13]

I apologise for the extremely crude crash course, but it was necessary in order to better explain the foundations of my cosmopsychistic position. The American theoretical physicist Geoffrey F. Chew explained the existential philosophical implication of the bootstrap theory as follows:

> *"Taken to its extreme, the bootstrap conjecture implies **that the existence of consciousness, together with all other aspects of nature, is necessary for the self-consistency of the whole.**"*[14]

In other words, consciousness would no longer be seen as an "external" aspect of our existence, not as something strictly separable and isolated from matter, but as an indispensable component of the whole of cosmic existence. And it goes even further: this "bootstrap assumption", as Chew called it, was the reason for me to see consciousness not as a consequence of material existence but, conversely, consciousness as an indispensable basic prerequisite for the observable and measurable phenomenon called matter. But this will be explained in more detail below.

Returning to the history of physics, one of its most important innovations in the 20th century was the realisation of the connection between the theory of the electromagnetic field and quantum theory. In a nutshell: photons are regarded as particles or can manifest themselves in this form, but they can also be described as electromagnetic waves, or as a wave field. These fields are in oscillation, in vibration. Both ideas were then brought together to form the concept of the quantum field: it can take the form of quanta - energy packets with a "probability of existence" - or particles. The contrast of classical Newtonian, mechanistic physics between "solid" particles and the space through which they move was thus resolved - the quantum field is regarded as the fundamental level of reality, a separation in the classical sense is no longer possible. This quantum field permeates the entire space, i.e. the entire universe. One could also say that "particles" are "condensates" in this wave field, i.e. "condensed energy", so to speak. Albert Einstein once expressed this as follows:

> *"We can therefore regard matter as consisting of regions of*

---

[13] Fritjof Capra, "The Tao of Physics. An Exploration of the Parallels between Modern Physics and Eastern Mysticism", Flamingo, 1982, p. 327
[14] Geoffrey Foucar Chew, "Bootstrap: A Scientific Idea", „Science", 1963, vol. 161, issue 3843, p. 763.

> *space in which the field is extremely intense ... In this new kind of physics, there is no place for the field and matter, because the field is the only reality."*[15]

Euclidean geometry, which formed the basis of the concept of space in classical physics and also had a major influence on Newton's mechanistic view of the world, was still considered a compulsory subject in schools until the beginning of the 20th century and was only overcome by Einstein's relativistic concepts. In Far Eastern philosophy, on the other hand, it was pointed out long before the emergence of relativity and quantum physics in the West that these were not real properties of nature, but constructs of human thought.

> *"It must be clear that space is nothing other than a mode of particularisation and that it has no real existence of its own.... Space exists only in relation to our particularising consciousness."*[16]

As I will be using quantum physics theories as the basis for my own speculations on the philosophy of nature and existence in this book, it is important to point out that these are often not definitive research findings, but often **preliminary** ones that are the subject of extensive debate among experts - not only in the natural sciences, but also in the philosophy of science. Many popular science publications and the press occasionally contain texts that unfortunately give the false impression that some of the supposedly central findings of quantum physics are virtually irrefutable, that they are established facts. Not at all. To give just one example: The famous "Copenhagen interpretation" is still the subject of intense debate to this day and is constantly being "shaken up" by new research findings.

> *"We have reached the end of our journey into the depths of matter. We have searched for solid ground and found none. The deeper we penetrate, the more restless the universe becomes,*

---

[15] A. Einstein, cite op. M. Capek, "The Philosophical Impact of Contemporary Physics", D. Van Nostrand, Princeton, New Jersey, 1961, p. 319.

[16] Ashvaghoosha, "The Awakening of Faith", translated by D.T. Suzuki, Open Court Chicago, 1900, p. 107.

> the more vague and cloudy. [...] There is no fixed place in the universe: **everything rushes and vibrates in a wild dance.**"[17]

But what should modern quantum physics have to do with questions of "mind" and "consciousness"? Scientists and philosophers of science turned to quantum physics because they realised that the methods of classical physics - a purely materialistic-mechanistic approach - simply could not convincingly explain how consciousness and thought can emerge from a complex interconnected neuronal system such as the human brain. Although there have been several attempts to explain consciousness on a purely material level, there have regularly been "gaps" in the explanations and contradictions that could not be eliminated with the "home remedies" of purely materialistic science. Of course, this does not imply that there can never be such an explanation; there is no reason to immediately surrender completely to metaphysical ideas such as those found in ancient mystery schools and religions. Here is another example:

> "The highest expression of matter is mind, which occupies the middle distance between activity on the one hand and motionlessness on the other. The mind of man is hypothetically conceived as consisting of two parts: the lower mind, which is connected with the demiurgic sphere of Jupiter, and the higher mind, which ascends to the substance of the divine power of Kronos and is similar to it. These two phases of mind are the mortal and immortal mind of Eastern philosophy. **The mortal mind is hopelessly entangled in the illusions of sense and substance, but immortal or divine mind, transcending these unrealities, is one with truth and light.** [...] Since intelligence is the highest manifestation of matter, it is logically the lowest manifestation of consciousness, or spirit, and Jupiter (or the personal ego) is confined within the substances of mortal mind, where it controls its world through what man likes to call intellect. The Jupiterian intellect, at any rate, is that which looks outward or towards the illusions of manifested existence, whereas the higher or spiritual mind (which lies dormant in most individuals) is that superior faculty which is able to think inward or into the depths of the self; in other words, which is

---

[17] Max Born, Nobel laureate in physics, from the postscript of his book "The Restless Universe", Springer, 1969, p. 166, my emphasis.

> *able to turn towards the substance of reality and look it in the face. Thus the mind could be compared to the two-faced Roman god Janus. With one face this god looks outwards, towards the world, and with the other inwards, towards the sanctuary in which he is encased. [...]* **The objective or mortal mind continually emphasises to the individual the supreme importance of physical phenomena; the subjective, or immortal mind - given the opportunity to express itself - combats this material instinct by intensifying attention to that which transcends the limitations of physical perceptions."**[18]

Consciousness research is quite simply an open field of research in which there are still no definitive answers. The assertion that the entire phenomenology of the human mind can be explained exclusively in physicalist-reductionist or materialist terms at the present time - given the current state of knowledge in research - cannot be upheld.

> *"It is widely accepted that consciousness or, more generally, mental activity is in some way related to the behaviour of the material brain. Since quantum theory is the most fundamental theory of matter currently available, it is legitimate to ask whether quantum theory can help us understand consciousness. [...] There are three basic types of corresponding approaches: (1) consciousness is a manifestation of quantum processes in the brain, (2) quantum concepts are used to understand consciousness without reference to brain activity, and (3) matter and consciousness are viewed as dual aspects of a single underlying reality."*[19]

Information theory is now also included:

> *"The science of consciousness has made great progress by focusing on the behavioural and neural correlates of experience. While these correlates are important for making progress, they are not sufficient to understand even basic facts, such as why the cerebral cortex gives rise to consciousness but the cerebellum does not, even though it has even more neurons and appears to be just as complicated. Moreover, correlates are of little help in many cases where we want to know whether consciousness is present: in patients with a few remaining*

---

[18] Manly P. Hall, „Lectures on Ancient Philosophy", Tarcher/Penguin, 2005, p. 18, my emphases.
[19] Harald Atmanspacher, Edward N. Zalta, „Quantum Approaches to Consciousness", https://plato.stanford.edu/entries/qt-consciousness/

> *islands of functioning cerebral cortex, in premature babies, in non-mammalian species, and in machines that rapidly outperform humans in driving, recognising faces and objects, and answering difficult questions. To address these questions, we need not only more data, but also a theory of consciousness - a theory that explains what experience is and what kinds of physical systems can have it. Integrated Information Theory (IIT) takes experience itself as its starting point (...)."*[20]

The quote makes a justified accusation: If the complexity of a particular system such as the human brain alone were a sufficient condition for the occurrence of consciousness or the sole cause, then modern high-performance computers, so-called "supercomputers", should have long since shown signs of consciousness due to their very high degree of interconnectedness at circuit level. So the material structure alone does not do it. If it is possibly the electrical currents themselves that cause phenomena such as consciousness and thinking on the basis of the material foundation - i.e. the complex neuronal "circuits" in the human brain - the question would again have to be asked whether this bioelectricity perhaps has "intrinsic", i.e. inherent properties that are peculiar to it, which must be included in the explanation. And at this point at the latest, quantum physics can no longer be ignored.

> *"As a result of the Cartesian* [based on the French philosopher René Descartes, author's note] *separation, most people are aware of themselves as isolated egos existing "inside" their bodies. The mind has been separated from the body and entrusted with the futile task of controlling it, creating an apparent conflict between the conscious will and the involuntary instincts. Each individual has been further divided into a large number of separate compartments according to his activities, talents, feelings, beliefs, etc., which are involved in endless conflicts that create constant metaphysical confusion and frustration."*[21]

In order to approach a possible answer to the question of why the belief in a "God" and the theory of evolution can be compatible and why no contradictions need arise

---

[20] Giulio Tononi, Christof Koch, „Consciousness: here, there and everywhere?", https://royalsocietypublishing.org/doi/10.1098/rstb.2014.0167
[21] Fritjof Capra, "The Tao of Physics. An Exploration of the Parallels between Modern Physics and Eastern Mysticism", Flamingo, 1982, p. 28.

from this, we must of course go back to the beginning: How did this world, this universe, actually come into being? What theories have been developed so far? And how credible are they? Furthermore: Why should the question of the origin of the cosmos have anything at all to do with "God", "spirit" or consciousness? Can't it all simply be explained within the framework of materialism, and if not, why not?

This brings us back to the question of the plausibility of the Big Bang theory. Is there really nothing to criticise about it? Is the mere fact that it has prevailed in such a way that it is still presented as the dominant theory, especially in the popular scientific press, and is used in everyday understanding as an explanation for the origin of the cosmos, proof of its correctness? Of course not.

What bothers me about this theory? According to this theory, the universe is said to have emerged from a so-called singularity, an *infinitely* small point. According to the current state of physics and astrophysics, the density of matter and energy at this point is said to have been *infinite* and all spatial distances are said to have been zero. So, in principle, **non-existent**. Because that is exactly what this concept boils down to: infinitely small means specifically: **non-existent**. Otherwise it would be a finite smallness - let's say, for example, a space of 10^{-18} m^3. In other words, a space with a volume of one attocubic metre - one trillionth of a cubic metre. But infinite means infinite: "infinitely small space" means that it does not exist at all (we wanted to pay attention to the accuracy of the terms). The Big Bang theory, in turn, led to the theory of an expanding universe - in other words, a space that has continued to expand over time and will continue to do so in the future. The common idea is that due to this continuous expansion, the cosmos will eventually die an „entropic death". Or it will contract/collapse and return into a singularity – maybe to give birth to another cosmos.

It is no coincidence that physicists have been arguing for a long time that the laws of nature as we know them have been suspended at this point in our ideas about the origin of the universe. No wonder, because the assumption of such a singularity actually contradicts some aspects of the laws of nature as we know them. According to these laws, matter can be compressed up to a certain point, **but not infinitely**. At such a point, the gravitational forces would theoretically be so enormous - namely *infinitely large* - that nothing would be able to explode outwards and form a cosmos, but could at most continue to implode. Since it is supposed to have been an "infinitely" small and dense singularity, it could only have imploded "infinitely" further. Such an idea would of course be absurd for the human imagination: how could an infinitely small, i.e. actually **non-existent** and dense point **continue to implode**? You can ask the brightest minds on this planet: they will all fail at precisely this point in the human conception of how the universe should have come into being. There are now several more or less plausible alternative theories. A brief overview will suffice here:

> „An alternative theory is the steady-state universe. An early rival to the Big Bang theory, steady state postulates the continuous creation of matter throughout the universe to explain its apparent expansion, according to NASA Cosmic Times. This type of universe would be infinite, without beginning or end. However, a mountain of evidence found since the mid-1960s shows that this theory is incorrect. Another alternative is the theory of eternal inflation. After the Big Bang, the universe expanded rapidly during a short period known as inflation. The theory of eternal inflation assumes that inflation never stopped and has continued for an infinite amount of time. Somewhere, even now, new universes are being created in a vast complex called the multiverse. In these many universes, different physical laws could apply. [...] The implications of quantum gravity and string theory point to a universe that is not in reality as it appears to the human observer. For example, it could be a flat hologram projected onto the surface of a sphere. Or it could be a completely digital simulation running on a giant computer."[22]

From my point of view, the idea of a universe that runs as a simulation on a computer is unintentionally funny, because it would of course immediately throw us back into an endless spiral of questions: Who or what created this computer? Who wrote the software that simulates our cosmos? And who created the programmer who wrote this software? Etc. etc. - ad infinitum.

And there are other problems too. For example, the fact of a "surplus" of matter: there is more matter in this universe than antimatter, although this should not be the case according to the conventional Big Bang theory. In other words, matter and antimatter should have been formed in equal proportions, which would have led to them cancelling each other out, resulting in an "empty" cosmos. Another problem is that the Big Bang theory cannot explain the strange fact that more gravity has been observed in the universe than there is matter to account for these gravitational forces. This is why other theories have also been developed over time, such as the one according to which our cosmos could have emerged from a previous, so-called "metaverse". Ervin Laszlo, Professor of Philosophy and Systems Theory, writes about this:

> "A number of physical cosmologies offer quantitatively elaborated explanations of how our universe could have come

---

[22] Karl Tate, „Alternatives to the Big Bang Theory", https://www.space.com/24781-big-bang-theory-alternatives-infographic.html

> *into being within the metaverse. Such cosmologies promise to solve the puzzles posed by the coherence of this universe, including the intriguing good fortune that its physical constants are so finely tuned that we can be here to ask questions about them. In a »one-shot«, one-cycle universe, there is no credible explanation for this, because there the vacuum fluctuations that set the parameters of the nascent universe would have been chosen at random: There was »nothing« that could have influenced the randomness of this selection. But it is **astronomically improbable** that a random selection from all possible fluctuations in the chaos of a turbulent primordial vacuum could have led to a universe in which living organisms and other complex and coherent phenomena emerge and evolve - or even to a universe in which there is a significant excess of matter over antimatter."*[23]

This concept of a "metaverse" - understood in the sense of a "quantum vacuum" or "energy field" that is possibly infinite and from which not just one, but in principle an infinite number of universes could have emerged in cyclical succession - will be encountered again in a brief discussion of the cosmogony of the Kabbalah.

> *"Most physicists no longer believe that the universe was created by the explosion of a single point and that time and space were created at that moment. Creation and evolution are not mutually exclusive concepts."*[24]

This brings us to another important question that also eludes the human mind. This is because we humans, as dimensional beings, cannot imagine complete dimensionlessness. Our entire perception, our entire thinking is characterised by concepts of space and time. We are therefore "condemned" to think in terms of finite causal chains - of cause and effect, of beginning and end. To put it rudely: we are finite causal chain-douchebags.

---

[23] Ervin Laszlo, "Science and the Akashic Field. An Integral Theory of Everything", Inner Traditions, Vermont, 2007, p. 39, my emphasis.
[24] C. Allan Boyles, "God and Quantum Physics", Wheatmark, 2021, p. 8

> *"Time, space and causality are like the glass through which one sees the absolute. In the absolute there is neither time, nor space, nor causality."*[25]

So if the universe was originally an "infinitely" small and dense point with a spatial extent of zero, we still have no choice but to ask: What was this singularity *in*? It can't have been there "just like that", can it? So: In which space, which "container" was this singularity located? And if this was the case and an infinitely dense singularity with extreme gravity existed in this space, then according to Einstein's theory of relativity, this gravitational force, which is unimaginably powerful for us, would have curved the surrounding space so strongly that, as already mentioned, nothing could have exploded outwards.

Or was there no such "container"? Did this singularity simply exist "for its own sake" or, to put it philosophically, "in and for itself"? Because, as I said, we cannot imagine a "nothing" of existence. Not to mention the fact that this would not make any sense, because nothing can come from nothing. Where is all the "matter" in the universe supposed to have come from? According to current astrophysical estimates, there are a staggering **one trillion galaxies** in the universe **visible to us so far**. That is **1000 billion**. As if this number alone were not breathtaking enough, each galaxy consists on average of **100-200 billion stars** (suns) - sometimes a little more, sometimes a little less - of which a certain percentage form solar systems, i.e. contain planets orbiting around their central star.

Would someone please be so kind as to explain to me how **one thousand billion galaxies** are supposed to arise from an **infinitely small point**, apart from which nothing else is supposed to have existed? In the meantime, there are even counts that assume **two to three trillion galaxies**. Excuse me, but the assertion that so much matter could emerge from an **infinitely** small point is really not very plausible. You don't need to be a trained nuclear physicist to realise this: It is known that atoms consist more or less of empty space, as the distances between the atomic nucleus and electron shells are very large, but even if you consider maximum compression, the math doesn't add up. At least not in the light of current scientific knowledge. Or maybe it is? Things look different in the realm of quantum physics, but we'll get to that later.

---

[25] Swami Vivekananda, "Jnana Yoga", Advaita Ashram, Calcutta, India, 1972, p. 109

Questions of this kind naturally give rise to others: **How long** did the singularity remain in its dormant state before it is said to have exploded into a universe, **and why**? Was it possibly the result of a previous contracting universe that - like an imploding sun exploding into a supernova - condensed until the motion reversed to create a new universe - the one we know today? In Hinduism, there is indeed the idea of universes that expand and contract in successive, infinite cycles. And this brings us back briefly to the idea of the "metaverse" just presented.

What exactly is this "metaverse"? What does it consist of? How is it structured? There are indeed concepts according to which this metaverse could be a "quantum vacuum", a **non-local** energy and "information field" that provides the necessary basic conditions for the creation of **local**, i.e. spatiotemporally "materialised" universes. When we talk about information here, this term is not to be understood in the common, conventional sense, which is why some authors write the word in an unusual form: **in-formation**. This means: in formation, in development.

> *"What in-formation is: in-formation is a subtle [...] connection between things in different places in space and events at different times. Such connections are called »non-local« in the natural sciences and »transpersonal« in consciousness research. Information connects things (particles, atoms, molecules, organisms, ecologies, solar systems, entire galaxies, and the mind and consciousness associated with some of these things) regardless of how far apart they are and how much time has passed since the connections between them were made."*[26]

The universes that may have emerged from this metaverse and the life forms that develop in them would then not only be **carriers** of information (e.g. laws of nature), but would themselves already be **a physical expression of this information**, i.e. living, dynamic, evolutionary **"representations"** of these in-formations. In other words, the metaverse would have "informed" our universe with regard to its creation and development. Which of course immediately raises the question again: **Who or what then, please, created this metaverse?**

> *"So how were the initial conditions of the metaverse created - by **what** ... or is the question, by **whom**? This is the deepest and greatest mystery of all - the mystery of the origins of the*

---

[26] Ervin Laszlo, "Science and the Akashic Field. An Integral Theory of Everything", Inner Traditions, Vermont, 2007, p. 69.

universe-creating process itself. This greatest of all mysteries is »transempirical«, it cannot be solved by reasoning based on observation and experiment. One thing is clear, however: if it is already unlikely that our finely tuned universe originated in a randomly configured vacuum, it is even more unlikely that the parent universe, which gave rise to a series of progressively evolving local universes, originated in a random, non-informed state. The vacuum of the metaverse was not only such that one universe could emerge in it, but such that a whole series of universes could emerge. This can hardly have been a lucky fluke. In some way, the original vacuum must have already been in-formed. There must have been an original creative act, an act of »metaversal design«."[27]

To return to the singularity of the Big Bang theory: These are man-made concepts that are naturally inadequate to describe a phenomenon such as an "infinitely small and dense" singularity, because they **emerged from observations of dimensional existence**. Our concepts are therefore "saturated" with dimensionality, they are the mental and linguistic expression of it. One could also say - forgive me if I am getting ahead of myself again for a moment - that as "local" beings, whose entire thinking and thus also their language is indelibly shaped by "**locality**", i.e. **by existential concretisation that has already taken place**, we cannot have an immediate view of "**non-locality**". Such a concept is "above our paygrade", it is too abstract for us.

> "Fish baskets are used to catch fish; but when the fish are caught, people forget the baskets; snares are used to catch rabbits; but when the rabbits are caught, people forget the snares. Words are used to convey ideas; but when you have grasped the ideas, you forget the words."[28]

Which is why we find it so difficult to answer another well-known question from astrophysics: Is the universe a finite or infinite space?

---

[27] Ibid., pp. 86-87.
[28] Chuang Tzu, translated by James Legge, compiled by Clae Wakham, Ace Books, 1971, chapter 26.

> *"However, the galaxies are not racing away from a common centre. The galaxies appear to be rushing away from the point at which you are currently located. The explanation for this is that the galaxies are not rushing apart, but that the space between the galaxies is constantly expanding."*[29]

It is a grandiose irony: if we start from the assumption of a ***finite*** space, we end up in an ***infinite*** spiral of questions from which there is no escape and to which there is no conclusive answer. Because if this space - the universe - is finite, where is it located? What is it "attached" to, what is it "anchored" in? Well, then this space is located in an even larger space. An extremely unsatisfactory answer, because what in turn is this even larger space located in? Is it a cosmic Matryoshka doll? In this way we never come to a conclusion. But if we imagine the cosmos as ***infinite*** space, we cannot form a concrete idea of it due to the ***finite nature of our thinking and its concepts***; we cannot have an immediate, empirical view of infinity. This is also trivial, because in order to be able to do so, we would have to be able to put ourselves in the state of infinity - only then would we know "what it feels like" to be infinite.

Of course, this dilemma could be „resolved" by simply drawing an arbitrary line: Outside of a certain final, lowest cause, there is nothing. In the Jewish Kabbalah, one of the most spiritual and profound texts one can read in this respect, such a boundary is drawn, at least *hypothetically*:

> *„AIN SOPH was referred to by the Qabbalists as The Most Ancient of all the Ancients. It was always considered as sexless. Its symbol was a closed eye. While it may be truly said of AIN SOPH that to define It is to defile It, the Rabbis postulated certain theories regarding the manner in which AIN SOPH **projected creations out of Itself**, and they also assigned to this Absolute **Not-Being** certain symbols as being descriptive, in part at least, of Its powers. The nature of AIN SOPH they symbolize by a circle, itself emblematic of eternity. This hypothetical circle encloses a **dimensionless area of incomprehensible life**, and the circular boundary of this life is **abstract and measureless infinity**. According to this concept, God is not only a Center but also Area. **Centralization is the first step towards limitation**. Therefore, centers which form in the substances of AIN SOPH are finite because they are*

---

[29] C. Allan Boyles, "God and Quantum Physics", Wheatmark, 2021, p. 117

> *predestined to dissolution back into the Cause of themselves, while AIN SOPH Itself is infinite because it is the ultimate condition of all things. The circular shape given to AIN SOPH signifying that space is hypothetically enclosed within a great crystal-like globe, outside of which there is nothing, not even a vacuum. Within this globe – symbolic of AIN SOPH – creation and dissolution take place. Every element and principle that will ever be used in the eternities of Kosmic birth, growth, and decay is within the transparent substances of this intangible sphere.*[30]

However, such an arbitrary demarcation would of course attract accusations of being unscientific and arbitrary: Who defines the point at which one is no longer allowed to continue thinking and researching, but "must" accept a "final frontier" as simply given and unquestionable?

Interim status so far: As we humans cannot help but think in such dimensional terms and causal chains - cause and effect, beginning and end - we are unfortunately denied the idea that a finite space can "just be there". The same applies to infinity: we cannot comprehend that something infinite can simply exist without being justifiable and questionable.

> *"That's where we reach the great barrier of thought, because we begin to struggle with the concepts of time and space before they existed in our everyday experience. I feel as if I have suddenly driven into a great wall of fog in which the familiar world has disappeared."*[31]

No matter how you look at it: We can imagine neither a finite nor an infinite space as existing "just like that". We are also denied a concrete idea of higher dimensions above the three spatial ones we are familiar with.

However, they can be expressed on an abstract level - namely mathematically. And it can also be visualised, e.g. on a computer screen. So if a scientist wants to know

---

[30] Manly P. Hall, „The Secret Teachings of All Ages. An Encoclopedic Outline of Masonic, Hermetic, Qabbalistic and Rosicrucian Symbolical Philosophy, being an Interpretation of the Secret Teachings concealed within the Rituals, Allegories, and Mysteries of all Ages", H.S. Crocker Company, Inc., 1928, p. 360, my emphases.
[31] A.C.B. Lovell, "The Individual and the Universe", Oxford University Press, London, 1958, p. 93

what we humans would see if, for example, a four-dimensional cube (a hypercube) were to plunge into or traverse our three-dimensional space, this is perfectly possible - but the resulting image would of course **still be three-dimensional. We cannot have an immediate, concrete visual idea** of four-dimensionality or even higher dimensions.

> *"All this implies that man with his natural, temporal intelligence can only find and recognise shadows, images and phenomenal forms of realities in this world, which exist eternally and noumenally in the world of the spirit, for which his temporal abilities are currently closed."*[32]

For comparison, consider the well-known example of purely two-dimensional beings living on a sheet of paper. Their world is therefore a two-dimensional surface. What would they see if a three-dimensional sphere were to cross their 2D plane of existence? They would probably be startled: As if out of "nowhere", a tiny dot would suddenly appear in their world. This dot "magically" expands into a circle with a growing radius. At a certain point, this circle also "magically" reverses the direction of its expansion: the radius shrinks again to become a dot and finally disappears from their 2D-0world again. Of course, this had nothing to do with magic, and the dot did not come out of "nothing" - it was simply a body whose three-dimensionality the inhabitants of the 2D world could not have any concrete idea of in their perception, which is limited to two dimensions. At some point in the course of their civilisational development, however, a clever 2D scientist could come to the conclusion that it was simply a 3D sphere that penetrated their 2D space.

What speaks against the fact that it could be the same for us humans? What speaks against the idea that what is commonly called "God" could simply be a higher-dimensional consciousness in the form of energy that only presents itself **to us** as a three-dimensional existence, since we cannot visualise such structures in their **actual, natural form** through our senses and our three-dimensional imagination? Is there even a single piece of really convincing, hard evidence to date that the "singularity" referred to in the Big Bang theory is perhaps not just what we would see if a four-dimensional or even higher-dimensional structure "crossed" three-dimensionality - or created it in the first place - in order to "expand" a universe? Why would that be "crazy"?

---

[32] Walter Leslie Wilmshurst, "The Meaning of Masonry", William Rider & Son, 8, Paternoster Row & Percy Lund, Humphries & Co., 3, Amen Corners, 1922, republished from Orkos Press, 2014, p. 96.

> "All these many different substances in this universe are merely many different states of pressure. These have been produced by the interaction of the opposite motion between two opposite poles of rest, which emerge from the zero-point universe of the **knowing mind** to imitate the manifold ideas of the **thinking mind**."[33]

As far as the idea of a singularity is concerned, there are also interesting thoughts on this in the Qabbalah. There, "God" is conceived of as a kind of primordial spirit or a consciousness of infinite expansion and complete timelessness, the so-called "Ain", from which a first point of manifestation emerged through a kind of spiritual focussing from an infinite state of non-existence - one could also say, a **mere potentiality or probability of principally infinite possibilities of manifestation**. From this uppermost sphere, ten more emerged, all of which are contained in the first, primordial, uppermost sphere. This process is repeated four times, so that four spheres emerged from ten sub-spheres each. The lowest sphere is that of material existence, to which our cosmos and we ourselves also belong. This is illustrated by a circular diagram:

> On the accompanying circular chart, the concentric rings represent diagrammatically the forty rates of vibration (called by the Qabbalists Spheres) which emanate from AIN SOPH. The circle X 1 is the outer boundary of space. It circumscribes the area of AIN SOPH. The nature of AIN SOPH Itself is divided into three parts, represented by the spaces respectively between X 1 and X 2, X 2 and X 3, X 3 and A 1; thus:
>
> X 1 to X 2, אין , AIN, the vacuum of pure spirit.
>
> X 2 to X 3, אין סוף , AINSOPH, the Limitless and Boundless.
>
> X 3 to A 1, אין סוף אור , AINSOPHAUR, the Limitless Light.
>
> It should be borne in mind that in the beginning the Supreme Substance, AIN, alone permeated the area of the circle; the inner rings had not yet come into manifestation. As the Divine Essence **concentrated Itself**, the rings X 2 and X 3 became apprehensible, for AIN SOPH is a limitation of AIN, and AIN

---

[33] Walter Russell, „Eine neue Vorstellung vom Universum" [A new concept on the Universe], Genius Verlag, Bremen, 2019, p. 92, my translation.

> SOPH AUR, or Light, is a still greater limitation. Thus the nature of the Supreme One is considered to be threefold, and from this threefold nature the powers and elements of creation were reflected into the Abyss left by the **motion of AIN SOPH towards the center of Itself**. The continual motion of AIN SOPH towards the center of Itself resulted in the establishment of the dot in the circle. **The dot was called God**, as being the supreme **individualization of the Universal Essence**. Concerning this the Zohar says:
>
> "When the concealed of the Concealed wished to reveal Himself He first made a single point: the Infinite was entirely unknown, and diffused no light before this luminous point violently broke through into vision."
>
> The name of this point is **I AM**, called by the Hebrews Eheieh. The Qabbalists gave many names to this dot.[34]

This supreme, primordial consciousness is incomprehensible to man:

> „The Qabbalists conceive of the Supreme Deity as an Incomprehensible Principle to be discovered only through the process of eliminating, in order, all its cognizable attributes. That which remains – when every knowable thing has been removed – is AIN SOPH, the eternal state of Being. Although indefinable, the Absolute permeates all space. Abstract to the degree of inconceivability, AIN SOPH is the unconditioned state of all things. Substances, essences, and intelligences are manifested out of the inscrutability of AIN SOPH, but the Absolute itself **is without substance, essence, or intelligence**."[35]

The Kabbalists visualised the emergence of human consciousness in a similar way:

> „As the consciousness in the Kosmic Egg is withdrawn into a

---

[34] Manly P. Hall, „The Secret Teachings of All Ages. An Encoclopedic Outline of Masonic, Hermetic, Qabbalistic and Rosicrucian Symbolical Philosophy, being an Interpretation of the Secret Teachings concealed within the Rituals, Allegories, and Mysteries of all Ages", H.S. Crocker Company, Inc., 1928, p. 362, my emphases.
[35] Ibid., p. 360, my emphasis.

> *central point, which is then called God – the Supreme One – so the consciousness in the Auric Egg of man is concentrated, thereby causing the establishment of a point of consciousness called the Ego. As the universes in Nature are formed from powers latent in the Kosmic Egg, so everything used by man in all his incarnations throughout the kingdoms of Nature is drawn from the latent powers within his Auric Egg. Man never passes from this egg; it remains even after death."*[36]

One could also say that in the course of an evolutionary process of self-realisation or self-development, the "cosmic consciousness" created the possibility of being able to "take a look at itself" through the development of human consciousness. We will come back to this question later - what this human "auric egg", as it is known in Kabbalah, could have to do with a so-called "Bose-Einstein condensate".

Manly P. Hall, a Canadian High Grade Freemason, mystic, philosopher and author of over 150 writings, provided the following conception of God in his book "Lectures on Ancient Philosophy":

> *"The God of tomorrow stands forth in all its majesty of suns and moons and stars. Its extent is from space to space, and eternity alone confines it. Man, gazing into the eyes of man, beholds therein his Maker. His Creator sings to him with the voice of the wilderness, and descends upon him from the stars that spangle the heavens by night. This God is not hidden behind flowing draperies, nor are his ministers avenging angels.* **Unmoved by the passing of ages he contemplates the worlds that are his substance, and through his own mind in man seeks to probe the depth of his own reality.** *This Vast One has written his law in the heavens where they shall endure long after earthly codes have been erased from the memory of man. This God manifests his will in the endless progression and change by which things are moved from Then to Now and from Now to Then. This Universal Creator fears not man's effort to understand him; telescopes and microscopes may scan his features without offence. For what is the quest of knowledge but the God in man seeking the God in All? God is; man is. Therefore, man is God and God is man. But before man may consciously enter into his divinity he must gaze upon himself in the All without fear and recognize himself in the All gazing*

---

[36] Ibid., S. 361

> *back without hate. Steadfastly and unafraid, **the rational soul thus gazes upon those glorious beings whose radiant natures are that mystical light which is the life of the beholding soul.***"[37]

This would essentially correspond to the perspective of "panpsychism", which views the mind or consciousness as a fundamental level of reality: In contrast to the explanatory models of physicalism, especially the strictly reductionist physicalism or materialism, according to which mental phenomena have a physical/material cause and are invariably attributable to it, panpsychism assumes that everything in the universe is an expression of consciousness. Within panpsychism itself there are many different sub-currents, which are beyond the scope of this book. After all, my aim here is to present it in a way that is as easy to understand as possible, so that even readers with no scientific or philosophical training can comprehend it. If you want to delve deeper into the subject, I recommend reading the article on panpsychism in the Stanford Encyclopedia of Philosophy.[38] A brief excerpt may suffice to illustrate the problem:

> *"There is a second prominent argument in favour of panpsychism that has nothing to do with the need to explain human consciousness; rather, it proceeds from a certain gap in the picture of the world that we get from the physical sciences. This argument has its roots in Leibniz, Schopenhauer, [Bertrand] Russell (1927) and [Alfred North] Whitehead (1933 [1967]) [...] In the public mind, physics is on its way to giving us a complete explanation of the fundamental nature of the material world. It seems almost tautological that »physics« is the true theory of the »physical« and that we should therefore turn to physics to understand the complete nature of space, time and matter. However, this commonly held view comes under pressure when we reflect on the rigorous vocabulary with which physical theories are formulated. A pivotal moment in the scientific revolution was Galileo's declaration that the book of the universe is written in the language of mathematics; since then, mathematics has been the language of physics. The vocabulary of physics is arguably not exclusively mathematical, as it contains causal or nomic concepts such as the concept of natural law. But the*

---

[37] Manly P. Hall, "Lectures on Ancient Philosophy", Philosophical Research Society, 1985, Jeremy T. Archer Edition, Penguin Books 2005, pp. 238-239, my emphases.
[38] Philip Goff (William Seager, Sean Allen-Hermanson), "Panpsychism", https://plato.stanford.edu/entries/panpsychism/

> kind of qualitative concepts that we find in the Aristotelian characterisation of the universe are completely absent from modern physics. Physical theories are formulated in a purely mathematical-nomical vocabulary. The problem is that it is not clear whether such a strict vocabulary can fully capture concrete reality even in principle. A mathematical description of a situation abstracts from concrete reality; a mathematical model in economics, for example, abstracts from what is bought or sold, or from the nature of labour. And nomic predicates can only express information about how physical entities will behave. That's fine if we want to predict how electrons will behave, for example. But intuitively, there must also be an intrinsic nature of an electron; there must be an answer to the question »How is the electron in and of itself?«. And this question apparently cannot be answered by describing how electrons should behave."[39]

Although the concept of panpsychism only emerged relatively late in intellectual history - its coining is attributed to the Venetian Renaissance philosopher Francesco Patrizi da Cherso (1529 - 1597), i.e. in the 16th century - there were already views that could be described as panpsychist much earlier.

> "Mystics and sages have long claimed that at the roots of reality there is an interconnected cosmic field that preserves and transmits information, a field known as the Akashic Record. Recent discoveries in vacuum physics show that this Akashic field is real and has its counterpart in the zero-point field of science, which underlies space itself. This field consists of a subtle sea of fluctuating energies from which all things emerge: Atoms and galaxies, stars and planets, living beings and even consciousness. This zero-point Akasha field is the constant, ongoing memory of the universe."[40]

The Greek philosopher Anaxagoras (499 - 428 BC), who can be categorised as an

---

[39] Philip Goff, William Seager, Sean Allen-Hermanson, „Panpsychism", https://plato.stanford.edu/entries/panpsychism/

[40] Ervin Laszlo, "Science and the Akashic Field. An Integral Theory of Everything", Inner Traditions, Vermont, 2007, from the back cover text.

atomist, described "God" as an infinite, self-moving spirit. This divine, infinite and incorporeal spirit was the effective cause of all things. Matter is infinite and consists of similar parts, from which everything is made by the divine spirit according to the will of this "creator" in order to give form and substance to the formless. These and similar views can be found throughout the history of philosophy. A small selection may suffice:

To name another Greek philosopher, Xenophanes (570 - 480 BC) declared that God is "a single and incorporeal God, round in substance and form, in no way resembling man; that he is all seeing and all hearing, but does not breathe; that he is everything, mind and wisdom, not generating, but eternal, immovable, unchanging and rational".[41] Xenophanes held the view that all existing things are eternal in terms of their ideal substance, that the world (by which he means the universe) has neither beginning nor end and that everything that is created is subject to transience.

Let's stay in ancient Greece for a moment: the philosopher Plato (428/427 - 348/347 BC) divided existence into three main categories:

> " According to Plato, the One is the term most suitable for defining the Absolute, since the whole precedes the parts and diversity is dependent on unity, but unity not on diversity. The One, moreover, is before being, for to be is an attribute or condition of the One. Platonic philosophy is based upon the postulation of three orders of being: that which moves unmoved, that which is self-moved, and that which is moved. That which is immovable but moves is anterior to that which is self-moved, which likewise is anterior to that which it moves. That in which motion is inherent cannot be separated from its motive power; it is therefore incapable of dissolution. Of such nature are the immortals. That which has motion imparted to it from another can be separated from the source of its an animating principle; it is therefore subject to dissolution. Of such nature are mortal beings. Superior to both the mortals and the immortals is that condition which continually moves yet itself is unmoved. To this constitution the power of abidance is inherent; it is therefore the Divine Permanence upon which all things are established. Being nobler even than self-motion, the unmoved Mover is the first of all dignities. The

---

[41] Manly P. Hall, „The Secret Teachings of All Ages. An Encoclopedic Outline of Masonic, Hermetic, Qabbalistic and Rosicrucian Symbolical Philosophy, being an Interpretation of the Secret Teachings concealed within the Rituals, Allegories, and Mysteries of all Ages", H.S. Crocker Company, Inc., 1928, p. 19-20.

> *Platonic discipline was founded upon the theory that learning is really reminiscence, or the bringing into objectivity of knowledge formerly acquired by the soul in a previous state of existence."*[42]

Well, isn't that a contradiction? "... the state that is constantly moving but is itself motionless"? Perhaps the following formulation would be better: the state from which all movement arises, but which itself is motionless. We will come back to this when it comes to the ideas that a certain Walter Russell developed about the universe.

---

[42] Ibid., pp. 17-18

> "In the same way, the IDEA that appears in matter is not in matter. The IDEA is never created. IDEA is a quality of the spirit. IDEA never leaves the omniscient light of consciousness. IDEA is only imitated by matter in motion. [...] All knowledge, all energy and all methods of creating any thing are qualities of consciousness alone. In matter, which is movement, there is no knowledge, no energy, no life, no truth, no intelligence, no materiality and no thinking."[43]

Dutch philosopher Baruch de Spinoza (1632 - 1677) conceived of God as a substance that exists absolutely in and for itself, which

> "... needs no other conception than itself to make it complete and intelligible. For Spinoza, the nature of this being was comprehensible only through its attributes, which are extension and thought: These combine to **form an infinite variety of aspects or modes**. The mind of man is **one of the modes of infinite thought**; the body of man is one of the modes of infinite extension. Through reason, man is able to rise above the illusory world of the senses and find eternal rest in perfect union with the divine essence. Spinoza, it is said, stripped God of all personality and equated the Godhead with the universe."[44]

I would respectfully disagree with Spinoza on one point: That "God" could be a substance or entity in need of no conception other than itself. For if that were the case, one would have to ask: If this substance or entity, if this "spirit" or "consciousness" exists in a self-sufficient form - why then appear as a creator at all? Why then create something? If I am completely sufficient for myself and exist as an absolutely self-sufficient state - why then set myself in motion and become creatively active at all? I will explore this question in more detail shortly. For now, I will leave it at another kabbalistic idea as a hint:

> "In the process of creation the diffused life of AIN SOPH retires from the circumference to the center of the circle and

---

[43] Walter Russell, „Eine neue Vorstellung vom Universum" [A new concept on the Universe], Genius Verlag, Bremen, 2019, p. 30, my translation.

[44] Manly P. Hall, Ibid., p. 31, my emphases.

> *establishes a point, which is the first manifesting One – the primitive limitation of the all-pervading O. When the Divine Essence thus retires from the circular boundary to the center, It leaves behind the Abyss, or, as the Qabbalists term it, the Great Privation. Thus, in AIN SOPH is established **a twofold condition where previously had existed but one.** The first condition is the central point – the primitive objectified radiance of the eternal, subjectified life. About this radiance is darkness caused by the deprivation of the life which is drawn to the center to create the first point, or universal germ. The universal AIN SOPH, therefore, no longer shines through space, but rather upon space from an established first point. Isaac Myer describes this process as follows:*
>
> *»The Ain Soph at first was filling All and then made an absolute concentration into Itself which produced the Abyss, Deep, or Space (...)«* "[45]

Interestingly, in Indian Hinduism there are very similar, millennia-old ideas about the "unity of reality", which is said to have emerged from a kind of "world soul", the "Brahman":

> *"The basis of Krishna's [an Indian deity, author's note] spiritual teaching, as well as of Hinduism as a whole, is the idea that the multitude of things and events around us are **but different manifestations of the same ultimate reality.** This reality, called Brahman, is the unifying concept that gives Hinduism its essentially monistic character, despite the worship of numerous gods and goddesses. Brahman, the ultimate reality, is understood as the "soul" or inner essence of all things. It is **infinite and beyond all concepts; it cannot be grasped by the intellect or adequately described in words**: Brahman, beginningless, sublime: beyond that which is and beyond that which is not. »Incomprehensible is this supreme soul, unlimited, unborn, unfathomable, unthinkable.«* "[46]

I'm sure readers will have noticed the obvious parallels between these Hindu and Kabbalistic views: A kind of infinite consciousness or "spirit" from which all existence emerged - everything we think of as material, substantial reality. However,

---

[45] Manly P. Hall, Ibid., p. 360-361, my emphasis.
[46] Fritjof Capra, "The Tao of Physics. An Exploration of the Parallels between Modern Physics and Eastern Mysticism", Flamingo, 1982, S. 99, my emphases.

this consciousness is so abstract - "... beyond that which is and beyond that which is not ..." (one could say, analogous to the infinite and eternal "Ain" of the Kabbalah) - that no polarisation into opposing states has yet taken place (i.e. no "Ain Soph" has yet emerged from the "Ain"). Logically, it cannot be grasped by human thought, caught up in dimensionality.

The same applies - albeit with some differences - to Buddhism, especially to the teachings of the so-called *Avatamsaka Sutra*:

> "The central theme of the Avatamsaka is the unity and interrelatedness of all things and events; a concept that is not only the essence of the Eastern worldview, but also **one of the fundamental elements of the worldview that emerges from modern physics**. It will thus be seen that the Avatamsaka Sutra, this ancient religious text, has the most striking parallels with the models and theories of modern physics."[47]

And such ideas can also be found in Chinese Taoism:

> "The Chinese, like the Indians, believed that there is an ultimate reality that underlies and unifies the manifold things and events that we observe [...] They called this reality the Tao, which originally meant »the way«. It is the way or process of the universe, the order of nature. [...] In its original cosmic meaning, the Tao is the **ultimate, indefinable reality** and as such is the equivalent of the Hindu Brahman and the Buddhist Dharmakaya. However, it differs from these Indian concepts in its inherently dynamic quality, which, according to the Chinese view, constitutes the essence of the universe. The Tao is the cosmic process in which all things are involved; the world is seen as a constant flow and change."[48]

German philosopher Gottfried Wilhelm von Leibniz (1646-1716) criticised Descartes' theory of extension on the basis of the criteria of "sufficient reason" developed by him and came to the conclusion that "substance" (matter) itself had an inherent force in the form of an incalculable number of separate and all-sufficient units. Reduced to its ultimate particles, matter ceased to exist as a substantial body and was dissolved into a mass of immaterial ideas or metaphysical units of force, for which Leibniz introduced the term "monad".

---

[47] Fritjof Capra, "The Tao of Physics. An Exploration of the Parallels between Modern Physics and Eastern Mysticism", Flamingo, 1982, p. 112, my emphasis.
[48] Ibid., 116-117, my emphasis.

> *"Matter is not made up of matter! [...] At the end of all the splitting up of matter, something remains **that is more akin to the spiritual** - holistic, open, alive: Potentiality, **the possibility of realisation**. Matter is the slag of this spirituality - decomposable, definable, determined: Reality."*[49]

According to him, the universe therefore consists of an infinite number of separate monadic units that unfold spontaneously through the "objectification" of their inherent active properties. All things are said to be composed of individual monads of different sizes or of aggregates of these bodies, which can exist as physical, emotional, mental or spiritual substances. According to Leibniz, God is the first and greatest monad, while the spirit of man is an "awakened" monad - in contrast to the lower realms, whose governing monadic powers are in a half-asleep state.

> *"Origen was born in Alexandria around 185 or 186. [...] The fundamental point in Origen's doctrine of God is that God is incorporeal. Origen claims that God cannot be seen by nature and argues that God **does not occupy a place.**"*[50]

The above quote about Origen will play a role in the discussion of the quantum physical principle of "non-locality".

Friedrich Wilhelm Joseph von Schelling (1775 - 1854), the successor to Johann Gottlieb Fichte (1762 - 1814) in the Chair of Philosophy at the University of Jena, recognised the necessity of certain objective realities and laid the foundations for a complete philosophical system with his "Theory of Identity". While Fichte regarded the self as the absolute,

> *"Schelling understood the infinite and eternal spirit as the all-pervading cause. The realisation of the absolute is made possible by intellectual intuition, which, as a higher or spiritual sense, **is able to distance itself from both the subject and the object.** Von Schelling understood Kant's categories of space and time as positive and negative respectively, and material existence as the result of the interaction of these two*

---

[49] Hans-Peter Dürr, physicist and essayist, former head of the Max Planck-Instituts of Physics, „Warum es ums Ganze geht", oekom-Verlag, pp. 86/87, my emphases.
[50] C. Allan Boyles, "God and Quantum Physics", Wheatmark, 2021, p. 127-128, my emphasis.

expressions. Von Schelling also held the view that the absolute, in its process of self-development, proceeds according to a law or rhythm consisting of three movements. The first, a reflective movement, **is the attempt of the infinite to embody itself in the finite.** *The second, that of subsumption, is the attempt of the absolute to return to the infinite after its entanglement in the finite. The third, that of reason, is the neutral point at which the two earlier movements merge."*[51]

French philosopher and writer Henri Bergson (1859 - 1941) advocated "a theory of mystical anti-intellectualism based on the premise of creative evolution. His rapid rise to popularity was due to his appeal to the finer feelings in human nature, which rebelled against the hopelessness and helplessness of materialistic science and realist philosophy. Bergson sees God as life that constantly struggles against the limitations of matter."[52]

English philosopher and sociologist Herbert Spencer (1820 - 1903), whose work "First Principles" made him one of the most famous philosophers of his time, defined "God" as follows:

"God is infinite intelligence, infinitely diversified through infinite time and infinite space, manifested through a myriad of constantly evolving individualities."[53]

Spencer emphasised the universality of the law of evolution, which he applied not only to form, but also to the intelligence behind the form. In every manifestation of being, he recognised the fundamental tendency of unfolding from simplicity to complexity, noting that reaching the point of equilibrium is always followed by a process of dissolution. According to Spencer, however, the dissolution only took place so that reintegration could take place on a higher level of being.

For Georg Wilhelm Friedrich Hegel (1770-1831), the logic of thought was of the utmost importance, namely as a property of the Absolute itself. Hegel envisioned God as a process of unfolding that never reaches a complete state, i.e. remains unfinished. In the same way, according to Hegel, thinking is without beginning or

---

[51] Manly P. Hall, „The Secret Teachings of All Ages. An Encoclopedic Outline of Masonic, Hermetic, Qabbalistic and Rosicrucian Symbolical Philosophy, being an Interpretation of the Secret Teachings concealed within the Rituals, Allegories, and Mysteries of all Ages", H.S. Crocker Company, Inc., 1928, p. 33, my emphases.

[52] Manly P. Hall, „The Secret Teachings of All Ages. An Encoclopedic Outline of Masonic, Hermetic, Qabbalistic and Rosicrucian Symbolical Philosophy, being an Interpretation of the Secret Teachings concealed within the Rituals, Allegories, and Mysteries of all Ages", H.S. Crocker Company, Inc., 1928, S. 19-20, meine Übersetzung und Hervorhebungen, p. 36

[53] Ibid., p. 37

end. He also believed *that all things owe their existence to their opposites and that all opposites are actually identical*. The only existence is the *relationship of the opposites to each other, through whose combinations new elements are created*. Since Hegel conceived of the divine spirit as an eternal, never-finished thought process, he thereby questioned the foundation of theism, and his philosophy limited immortality to the eternally flowing God alone. Consequently, *evolution for him was the never-ending flow of divine consciousness out of itself*. The whole of creation, although constantly in motion, never reaches a state other than that of endless flow.

> "We have prioritised self-insurance over integration, analysis over synthesis, rational knowledge over intuitive wisdom, science over religion, competition over cooperation, expansion over conservation, and so on. This one-sided development has now reached a most alarming stage: a crisis of social, ecological, moral and spiritual dimensions."[54]

It is very interesting to note that the ideas of several great thinkers and philosophers of intellectual history about existence, God and the universe described above were by no means "fresh" ideas that were only developed during their lifetime, but rather much earlier. It is sometimes difficult to determine the extent to which later philosophers had knowledge of the concepts of ancient mystery schools and religions and their cosmogonies and theistic belief systems; however, due to their great literacy and education in such matters, it can be assumed that at least some of them drew on older philosophies and were influenced by them in the development of their own ideas, not to mention the transmission of this knowledge through the centuries in the corresponding secret societies such as the Knights Templar, Rosicrucianism or Freemasonry.

As soon as this topic comes up, people quickly shake their heads, because it is considered "dubious"; however, the existence of such secret societies and lodges is just as much a historical fact as their existential philosophical and theistic systems, which have been passed down in literature, and the remarkable fact that many a philosopher known around the world today may have been an "initiate", i.e. "initiated" into the corresponding systems - such as Pythagoras or Plato, for example, who are said to have had contact with these schools while travelling in Egypt and

---

[54] Fritjof Capra, "The Tao of Physics. An Exploration of the Parallels between Modern Physics and Eastern Mysticism", Flamingo, 1982, p. 15.

were also initiated there. The long-lasting historical continuity in the transmission of this esoteric information alone is of course no proof of its accuracy - this is not even implied here, but merely a reference to the historical permanence of certain fundamental existential-philosophical-theistic ideas. Whether the historical persistence of these concepts is really only due to a long-lasting tradition of inadequate knowledge, as quite a few scientists today like to claim, or whether great thinkers of the past perhaps intuitively had the "right instinct", remains to be seen. However, the parallels between the ancient cosmogonies and the findings and theories of modern quantum physics that have already been recognised by many physicists - including Werner Heisenberg and Niels Bohr - are and remain remarkable.

> *"If the modern scientific world could even guess at the true depth of these philosophical conclusions of the ancients, it would realise that those who fabricated the structure of the Kabbalah possessed a knowledge of the heavenly plan comparable in every respect to that of the modern scholar."*[55]

We should also not forget the possibility that the "metaphysical verbiage" of earlier times, so often ridiculed by modern science, was perhaps not due to "primitive" knowledge at all, but merely to a language that did not yet have the modern, scientific-technical vocabulary at its disposal. In other words: perhaps the greatest philosophers of antiquity did not use rather flowery language or many linguistic images - metaphors, allegories and similes - because they were less developed or "dumber", but because the conceptual tools of modern science did not yet exist. It could be a mistake to make hasty judgements here.

In addition, there are different methods of investigation, different approaches: The natural sciences proceed empirically - through observation, analysis, evaluation of measurement data etc., they direct their gaze outwards, they observe the "outside world"; in the natural and existential philosophical-religious mystery traditions, on the other hand, the gaze is directed inwards, into one's own consciousness, so one concentrates on the "inner world" through meditation, for example.

In addition, there are different methods of investigation, different approaches: The natural sciences proceed empirically - through observation, analysis, evaluation of

---

[55] Manly P. Hall, „The Secret Teachings of All Ages. An Encoclopedic Outline of Masonic, Hermetic, Qabbalistic and Rosicrucian Symbolical Philosophy, being an Interpretation of the Secret Teachings concealed within the Rituals, Allegories, and Mysteries of all Ages", H.S. Crocker Company, Inc., 1928, p. 380.

measurement data etc., they direct their gaze **outwards**, they observe the "outside world"; in the natural and existential philosophical-religious mystery traditions, on the other hand, the gaze is directed **inwards**, into one's own consciousness, so one concentrates on the "inner world" through meditation, for example. Assuming that human consciousness is actually only an individualized partial expression of a larger "cosmic whole", i.e. a "fragment" or a "splinter" of a universal cosmic "light" (consciousness) that unfolds through space and time like through a "prism" in manifold "spectral colors" (manifestations) for the purpose of becoming self-aware - from this perspective it would only be logical if people can also gain correct insights **intuitively** - by looking into their own spiritual inner life - on the basis of this "spiritual participation" in the overall process, even if these are not formulated as precisely as in the natural sciences. I don't believe that these are **fundamentally** opposing methods, but **complementary** ones; I believe that both can complement each other in a very fruitful, productive way. Many great scientists sometimes followed their intuitions, their "hunches", which led them to important insights - which they then formulated in the form of mathematical theorems, for example, or confirmed through experiments.

> *"Modern physics has dramatically confirmed one of the fundamental ideas of Eastern mysticism: that all the concepts we use to describe nature are limited, that they are not features of reality, as we tend to believe, but creations of the mind; parts of the map, not of the territory."*[56]

As far as the ideas of God of the aforementioned philosophers are concerned, they can be found in a more or less similar form in the so-called "Hermetic teachings", which can be traced back to a mysterious figure called Hermes Mercurius Trismegistus or Hermes Trismegistos. His exact identity is shrouded in the darkness of history; he is also often referred to as a mystical artificial figure, a "divine figure", who is said to have emerged from a fusion of the Greek god Hermes with the Egyptian deity Thoth. The idea of a "Supreme Spirit" - a creative, organising mind as the cause of the universe and all existence within it - is not only found in Qabbalah, as we have already seen, or in Indian Hinduism, but also in ancient Asian philosophies and religions.

In his book "The Secret Teachings of All Ages", published in 1928, Manly Palmer Hall provided an outline of Hermetic cosmogony, which I have shortened as far as

---

[56] Fritjof Capra, "The Tao of Physics. An Exploration of the Parallels between Modern Physics and Eastern Mysticism", Flamingo, 1982, p. 178.

possible in order to concentrate the text on the points relevant to this book:

"*The Divine Pymander of Hermes Mercurius Trismegistus is one of the earliest Hermetic writings extant today. Although it is probably not in its original form, having been remodelled in the first centuries of the Christian era and mistranslated since then, this work undoubtedly contains many of the original concepts of the Hermetic cult. [...] The Vision is the most famous of all the Hermetic fragments and contains an exposition of Hermetic cosmogony and the secret sciences of the Egyptians concerning the culture and unfoldment of the human soul. For a time it was erroneously called »The Genesis of Enoch«, but this error has now been corrected. [...]*

As Hermes wandered in a rocky and solitary place, he devoted himself to meditation and prayer. Following the secret instructions of the temple, he gradually freed his higher consciousness from the shackles of his physical senses; and thus liberated, his divine nature revealed to him the secrets of the transcendental spheres. He beheld a terrible and terrifying figure. It was the Great Dragon, with wings stretching across the sky and light streaming from its body in all directions. (The Mysteries taught that universal life was personified as a dragon). The Great Dragon called Hermes by name and asked him why he was meditating on the World Mystery in this way. Shocked at the sight, Hermes prostrated himself before the dragon and asked him to reveal his identity. The great creature replied that he was Poimandres, the spirit of the universe, the creative intelligence and the absolute ruler of everything. [...] Hermes then asked Poimandres to reveal the nature of the universe [...].

Poimandres' shape immediately changed. Where he had been standing, there was a radiant and pulsating light. [...] His mind told Hermes **that the light was the form of the spiritual universe** *and that the swirling darkness that had engulfed it* **represented the material substance.** *[...] Then the voice of Poimandres was heard again, but his form was not revealed:*

»I, your God, am the light and the spirit that were before substance was separated from spirit and darkness from light. And the **Word** *that appeared like a pillar of flame out of the*

*darkness is the Son of God, born of the mystery of the Spirit. The name of this Word is **Reason**. Reason is the offspring of thought, and reason will separate the light from the darkness and establish the truth in the midst of the waters. Understand, O Hermes, and meditate deeply on the mystery. That which thou seest and hearest is not of the earth, but the word of God incarnate. Thus it is said that the divine light dwells in the midst of mortal darkness, and ignorance cannot separate them. The union of the Word and the Spirit produces that mystery which is called life. [...]«*

*At the word of the dragon, the heavens opened and the myriad powers of light revealed themselves, soaring through the cosmos on wings of streaming fire. [...] Hermes realised that what he saw was only revealed to him because Poimandres had spoken a word. **The word was reason**, and through the reason of the word the invisible things were revealed. The divine mind - the dragon - continued his speech:*

*»Before the visible universe came into being, its mould was cast. This mould was called the archetype, and this archetype was in the highest mind long before the process of creation began. When the highest mind beheld the archetypes, it was enamoured with its own thought; so it took the Word as a mighty hammer and dug out caves in **primordial space** and poured the form of the spheres into the archetypal mould, simultaneously sowing the seeds of living beings into the newly formed bodies. The darkness below us, which received the hammer of the Word, was moulded into an ordered universe. The elements separated into layers and each gave birth to living beings. The highest being - the mind - male and female, brought forth the Word; and the Word, hovering between light and darkness, was liberated by another mind, called the labourer, the builder, or the creator of things. [...]*

*In this way it was accomplished, O Hermes: the Word, moving like a breath through space, called forth the fire by the friction of its **movement**. [...] This will continue from an infinite beginning to an infinite end, for the beginning and the end **are in the same place and in the same state**. [...] Then the downward and unreasonable elements produced creatures without reason. Matter could not produce reason, for reason* *had risen from it. The air produced flying creatures and the waters those that swim. The earth brought forth strange four-footed and crawling animals [...]*

Man longed to penetrate the circumference of the circles and understand the mystery of the One who sat on the Eternal Fire. Already possessing all power, he bent down and peered through the seven harmonies and broke through the power of the circles, revealing himself to the nature below. Man, gazing into the depths, smiled, for he saw a shadow on the earth and an image reflected in the water, and this shadow and image were a reflection of himself. Man fell in love with his own shadow and wished to descend into it. Simultaneously with this desire, the intelligent thing united with the unreasonable image or form. [...]

Nature, seeing the descent, wrapped itself around the man it loved and the two mingled. For this reason the earthly man is composed. Within him is the heavenly man, immortal and beautiful; outside is nature, mortal and destructible. Thus suffering is the consequence of the immortal man falling in love with his shadow and giving up reality to dwell in the darkness of illusion; for since man is immortal, he has the power [...] also of life, light and the word - but since he is mortal, he is controlled by the rings of rulers - fate or destiny.«"[57]

I have emphasised the "primordial space" and the phrase "... because the beginning and the end are in the same place and in the same state" because we will encounter this idea again in a brief discussion of Bohm's "pre-space" and the aforementioned quantum physical concept of **non-locality**.

The crucial point is that the idea of a spirit or consciousness, a kind of energy - which generates a **movement** from within itself - one could also say, **begins to vibrate, to oscillate** - could represent an extremely important key to understanding the cause of existential concretisation (manifestation), e.g. in the form of a universe - a "localisation" from a "non-locality", to use a term from quantum physics again. Or, if you will, a "collapse of the probability function", namely through "self-

---

[57] Manly P. Hall, „The Secret Teachings of All Ages. An Encoclopedic Outline of Masonic, Hermetic, Qabbalistic and Rosicrucian Symbolical Philosophy, being an Interpretation of the Secret Teachings concealed within the Rituals, Allegories, and Mysteries of all Ages", H.S. Crocker Company, Inc., 1928, pp. 98-101, my emphases.

observation" or "self-measurement".

Incidentally, you may have wondered why a dragon was used as a symbol for these energies in the above hermetic creation narrative. Jeremy Narby writes about this in his book "The Cosmic Serpent. DNA and the Origins of Knowledge":

> "According to the Dictionary of Symbols, *the dragon represents »the union of two opposing principles«.*"[58]

To eloborate on this topic a bit further, according to the "Dictionary of Symbols" by J.E. Cirlot, dragons as symbolic animals take

> "within the cosmic order [...] an intermediate position between **the world of fully differentiated beings and the world of formless matter.** They may have been inspired by the discovery of skeletons of antediluvian animals, but also by certain creatures which, although natural, have an ambiguous appearance (carnivorous plants, sea urchins, flying fish, bats) and thus stand for **change and transformation, but also for purposeful evolution into new forms.**"[59]

At this point, one could return to the discussions surrounding panpsychism, i.e. the question of the possible cause of existence from spirit and consciousness in contrast to the reductionist-physicalist view, according to which everything has only a material basis as its cause. But more on this in a moment. Firstly, I would like to give another example of such existential-philosophical-speculative questions of mystical provenance, namely in the form of a name that some may have heard or read somewhere: the Masonic figure of "Hiram Abiff", around whom many interpretations entwine. For some, he is a real historical figure; others, however, such as Manly P. Hall quoted above, describe him as an ***allegorical figure*** who stands for the "divine spirit" that lies dormant in matter as its "moving" form of expression. Hall calls him "CHiram":

> "CHiram is called dead because the cosmic creative forces in the average individual are **limited in their manifestation to a purely physical - and correspondingly materialistic - expression.** [...] Just as the sunlight symbolically dies as it approaches the winter solstice, the physical world can be described as the winter solstice of the spirit. On reaching the winter solstice, the sun appears to stand still for three days and

---

[58] Jeremy Narby, „The Cosmic Serpent. DNA and the Origins of Knowledge", Weidenfeld & Nicolson, 1999, p. 83, my emphasis.
[59] J.E. Cirlot, "A Dictionary of Symbols", Routledge & Kegan Paul Ltd., 1962, p. 11, my emphases.

then, rolling away the stone of winter, begins its triumphal march northwards to the summer solstice. The state of ignorance can be compared with the winter solstice of philosophy, spiritual understanding with the summer solstice. From this point of view, the initiation into the Mysteries becomes the spring equinox of the spirit, when the CHiram in man passes from the realm of mortality to that of eternal life. The autumn equinox corresponds to the mythological Fall of Man, when the human spirit descended into the realms of Hades by immersing itself in the illusion of earthly existence."[60]

But to come back to the question: Is our human consciousness perhaps only a "fragment", a partial expression of a universal consciousness of existence in "I-form"? An ego form "shadowed" by sense-based perception and the fact that it is only a partial, finite realisation of it, which cannot fully grasp its source due to its limited perception and dimensional limitations? In the following pages, it will become clearer why I have asked these questions - when it comes to the question why people are actually able to recognise mathematical laws at all.

"God sleeps in the stone, breathes in the plant, dreams in the animal and awakens in man" - Indian wisdom

The question of the origin of existence, which we refer to as the universe, matter or nature, cannot be answered completely satisfactorily with the models of the natural sciences available to date. For the time being, therefore, there is nothing left to do but to practise speculative natural philosophy or existential philosophy with regard to the still "incomprehensible remainder". But where should we start? Firstly, a very simple and at first glance certainly seemingly silly example from the language of everyday life, with which I will cautiously approach the concept of "spirit" that I have in mind.

Imagine an elderly gentleman with a white beard meets you on the street, greets you in a friendly manner and says: "Hello, I'm a carpenter."

You would probably ask: "How am I supposed to know if that's true? Have you

---

[60] Manly P. Hall, „The Secret Teachings of All Ages. An Encoclopedic Outline of Masonic, Hermetic, Qabbalistic and Rosicrucian Symbolical Philosophy, being an Interpretation of the Secret Teachings concealed within the Rituals, Allegories, and Mysteries of all Ages", H.S. Crocker Company, Inc., 1928, p. 236, my emphasis.

created anything that identifies you as a carpenter? A table, a chair, a bookshelf, a bed - it doesn't matter, as long as it's made of wood?"

"No, I haven't," replies the old bearded man.

"Then you can't know and claim to be a carpenter!"

Nobody can know that they are a carpenter as long as they have never worked in this field. At most in their imagination, in their consciousness - but not in the form of a concrete manifestation of the "carpentry idea".

Analogue to this: A "creator" ***cannot know that he is one if he has never created anything.*** For this ***self-perception and self-definition***, he would need a ***point of reference through which he can first perceive himself as something*** - something that is not himself, so that a definition as "I am something else" can succeed. What's more, in order to be able to say "I am" at all. A frame of reference is therefore necessary. <u>And please don't forget:</u> I am only dealing with man-made and therefore inevitably very incomplete, inadequate terms here. Call it what you like: "God", "creator", "consciousness", "spirit", "energy field" - whatever makes you happy.

In very general terms: ***How is "something" supposed to know what "it" is if there is nothing else through which "it" can perceive itself as "something" in the first place?***

> *"Things receive their essence and their nature through interdependence and are nothing in themselves"*.[61]

As I mentioned above, I therefore do not entirely share Spinoza's conception of God, according to which this absolute spirit needs "no other conception than itself" in order to "make it complete and intelligible". If this were the case, the logical consequence would be the question of why "something" that is completely sufficient unto itself and possesses an absolute understanding of itself should create anything at all. Why then set itself "in motion" at all?

> *"Quantum cosmology, in which the relevant quantum system is the entire universe and there is therefore no observer outside*

---

[61] Nagarjuna, cit.op. T.R.V.Murti, "The Central Philosophy of Buddhism", Allen & Unwin, London, 1955, p. 138.

> the system that causes the collapse of the wave function during the measurement, is particularly problematic from an orthodox point of view."[62]

As far as the question of the nature and essence of consciousness is concerned, one can essentially take two positions: The materialist or psychist or panpsychist. Materialistic science, of course, says: Oh, these are just bioelectrical and -chemical processes in a complex neuronal network called the brain. This reductionist-physicalist attitude means that everything can be reduced to physical processes and material elements.

> "One could say that for Spinoza, physical science is a way of studying the psychology of God. There is nothing in nature that does not have a mental aspect - the correct assessment of matter itself reveals that it is the other side of a mentalistic coin."[63]

The complexity of this lump of matter called the brain in our heads has given us the ability not only to react to our environment, but also to act proactively - on our own initiative, i.e. on the basis of what is known as "free will". Incidentally, the latter - "free will" - has also been under discussion for a long time; articles appear in the press from time to time, e.g. by neurobiologists or philosophers, who more or less deny humans this ability. In any case, according to "materialistic science", there is nothing more to say - there is only matter, everything else would be metaphysical rubbish, unscientific superstition. Period! A questionable position, which is why I chose the quote from the American physicist and astronomer Stephen Barr as the opening quotation for this book:

> "What many people think is a conflict between religion and science is actually something else. It is a conflict between religion and materialism. **Materialism regards itself as scientific** and is often labelled »scientific materialism« even by its opponents, **but it has no legitimate claim to be part of science. Rather, it is a school of philosophy** characterised by the belief that nothing exists apart from matter, or, as

---

[62] Sheldon Goldstein, „Bohmian Mechanics", https://plato.stanford.edu/entries/qm-bohm/
[63] Philip Goff, William Seager, Sean Allen-Hermanson, „Panpsychism", https://plato.stanford.edu/entries/panpsychism/

Democritus put it, »atoms and nothingness«."

This physicalist or materialist-reductionist world view is opposed by the aforementioned panpsychism, the main features of which are briefly repeated here:

> *"The role of physics is to provide us with mathematical models that accurately predict the behaviour of matter. This is incredibly useful information; a comprehensive understanding of the causal structure of matter has enabled us to manipulate the natural world in all sorts of extraordinary ways, allowing us to build lasers and hairdryers and put men on the moon. We can explain the extraordinary success of physics by the fact that, since Galileo, it has focussed on this limited project of capturing the causal structure of matter, rather than speculating about the underlying nature of the material that has that structure.* **As philosophers, however, we may be interested in finding out what the actual nature of matter is, or at least having our best guess as to what it might be.** *[...] The panpsychist has a suggestion: the intrinsic nature of matter is, at least in part, consciousness. Assuming for the sake of discussion that electrons are fundamental constituents of reality, the panpsychist's proposal is as follows: Physics tells us how an electron behaves, but in and of itself the electron is essentially a thing that instantiates consciousness (presumably of a very fundamental sort). What can be said in favour of this proposal? First of all, it is not obvious that we have an alternative proposal, at least not at this stage. We learn about matter through its causal effect on our senses or on our measuring instruments; as Eddington (1928: 58-60; cited in Strawson 2006a) put it, "Our knowledge of the nature of objects treated of in physics consists entirely in the readings of pointers [on instrument scales] and other indicators". It is difficult to imagine how this indirect method of studying matter could provide insights into its actual nature. Derk Pereboom (2011) has suggested that future thinkers might arrive at a positive hypothesis about the intrinsic nature of matter through their imagination, and such a proposal might prove to have empirical or other theoretical support.* **However, the natural sciences, at least in their present form, have no use for speculation about the intrinsic nature of matter. We have a choice between the panpsychist proposal and the view that**

the intrinsic nature of matter is »we don't know what«."[64]

As far as the physicalist/materialist approach, also known as the "strong reduction" approach (reductionism), is concerned,

> *"much discussed counterarguments against the validity of such strongly reductionist approaches [...] are the qualia arguments, which emphasise the impossibility of physicalist accounts to adequately account for the quality of the subjective experience of a mental state, the "what it is like to be in that state" (Nagel 1974). [...] Another, less discussed counterargument is that the physical realm itself is not causally closed. Any solution to the fundamental equations of motion (whether experimental, numerical or analytical) requires the specification of boundary and initial conditions that are not given by the fundamental laws of nature (Primas 2002). This causal gap applies to both classical physics and quantum physics, where a fundamental indeterminacy due to collapse [of the wave function, author's note] poses an even greater challenge. A third class of counterarguments relates to the difficulties of incorporating notions of temporal presence and unknowability into a physical description (Franck 2004, 2008; Primas 2017)."*[65]

German physicist Max Planck (1858-1947), one of the founders of quantum physics, saw no major tensions between science and religion:

> *"Religion and natural science meet in the question of the existence and nature of a supreme power ruling over the world. Despite this agreement, however, a fundamental difference must be noted: For religious people, God is given directly and primarily. From him, from his omnipotent will springs all life and all events, in the physical as well as in the spiritual world. Even if he is not recognisable with the intellect, he is nevertheless directly grasped through the religious symbols in contemplation and places his holy message in the souls of those who entrust themselves to him in faith. In contrast, for the natural scientist the only primary given is the content of his sensory perception and the*

---

[64] Philip Goff, William Seager, Sean Allen-Hermanson, "Panpsychism", https://axelkra.us/panpsychismus-philip-goff-william-seager-sean-allen-hermanson/, my emphases.
[65] Ibid.

> measurements derived from it. So if both religion and natural science require faith in God in order to operate, then for some God stands at the beginning, for others at the end of all thought. For some it is the foundation, for others the crown of the structure of any worldview. Wherever and however far we look, we find no contradiction between religion and natural science, but rather complete agreement on the decisive points."[66]

His colleague Albert Einstein also discovered a religious component in the natural sciences:

> "You will hardly find a scientific mind that digs deeper without a peculiar religiosity. This religiosity, however, differs from that of the naive person. To the latter, God is a being whose care one hopes for, whose punishment one fears - a sublimated feeling of the kind of relationship a child has with its father - a being with whom one has, as it were, a personal relationship, however respectful this may be. The researcher, however, is imbued with the causality of all events. The future is no less necessary and determined to him than the past. For him, morality is not a divine but a purely human matter. **His religiosity lies in his enraptured amazement at the harmony of natural law, in which such superior reason is revealed that all meaningful human thought and organisation is a completely vain reflection in comparison.** This feeling is the leitmotif of his life and endeavours, insofar as it can rise above the bondage of selfish desires. Undoubtedly, this feeling is closely related to that which has filled the religiously creative natures of all times."[67]

It is therefore not about personalised concepts of God of the "old man with the white beard", which would indeed be grotesquely ridiculous. As mentioned at the beginning of this book, the question of a possible compatibility between the concept of "God" or creation and a gradually unfolding life (evolution) therefore depends on

---

[66] https://axelkra.us/wp-content/uploads/2023/04/Max-Planck_-_...denn-die-Materie-bestuende-ohne-de_haupt-nicht._.mp4, my translation.

[67] Albert Einstein, „Mein Weltbild" (My worldview), chapter „Die Religiosität der Forschung" (The religiosity of research), published by Carl Seelig, Ullstein Verlag, p.18, my emphasis.

the concrete conceptualisation of the concept of God and evolution. There is no reason to arbitrarily declare the two incompatible - especially not in view of the fact how little man still knows, despite all his scientific achievements, not only about the cosmos, but also about his own existence, consciousness and thinking.

From my point of view, it is therefore incredibly presumptuous and stems more from a narrow-minded view of the world to simply declare flatly that there can be no "God" in the sense of an organising, creative reason, intelligence or consciousness; that everything came about completely by chance, that no form-giving reason can be discovered behind it - that everything just came about "just like that" out of pure chaos. To this day, even the "smartest scientists" of the materialist or reductionist-physicalist faction **cannot say**, for example, what exactly electricity actually is; they can describe it approximately in terms of electrons, electrical charges and their flows between voltage poles, but *cannot* provide an exact explanation of its intrinsic nature; they still don't know what *exactly* gravity actually is; they are still puzzling over what "dark matter" in the cosmos is all about, to name just a few examples - but of course they want to know for sure that anyone who believes in a "higher reason" or intelligence, a creative consciousness or a "god" for all I care, is a superstitious nutcase or dreamer who is not to be taken seriously at all.

> *"This universe of matter in motion is a consciously conceived, consciousness-generating body. As such it is as much a product of the mind as a pair of shoes, a poem, a symphony or a tunnel through a mountain are the product of the mind by which they were conceived and which produced the movement that created them as bodies in form and matter."*[68]

For the sake of simplicity, I will continue to use the term "mind". This is also because "mind-matter dualism" is a long-established question in philosophy, which was given new scientific and experimental substance with the advent of quantum physics - in the form of "particle-wave dualism". Incidentally, this has also already been called into question: The "pilot wave" or "guide wave" model of the American quantum physicist and philosopher David Bohm (1917 - 1992) attempted to explain the results of the famous "double-slit experiment" by the fact that particles are carried by "guide waves". In this double-slit experiment, it was discovered that

---

[68] Walter Russell, „Eine neue Vorstellung vom Universum" (A new concept of the Universe), Genius Verlag Bremen, 2019, p. 29, my translation.

photons behave like waves and form interference patterns on the screen behind the double slit, but when an exact measurement of individual photons was attempted, they exhibited behaviour that was more indicative of particle character.

> *"The apparent contradiction between the particle and the wave picture was resolved in a completely unexpected way that challenged the foundation of the mechanistic worldview - the notion of the reality of matter. At the subatomic level, matter does not exist with certainty in certain places, but rather shows »tendencies to exist«, and atomic events do not occur with certainty at certain times and in certain ways, but rather show »tendencies to occur«. In the formalism of quantum theory, these tendencies are expressed as probabilities and associated with mathematical quantities that take the form of waves."*[69]

Without going into too much detail here, such experiments ultimately gave rise to the so-called "paradox of the intelligent observer", which I question because I consider the possibility that human consciousness and thought could have a non-local source and in their "collapsed" form - in the form of observation by an existentially concretised entity called "man" - could ultimately amount to a "self-observation" of a more universal consciousness in individually reduced partial aspects. My view of existence (in its most general term as a description of the entire cosmic existence) can thus be described as panpsychistic, or more precisely: as **constitutive cosmopsychism**; I consider the strictly materialistic approach, which seeks to attribute consciousness and thought exclusively to material causes, to be a dead end.

> *"Even stranger is the realisation that a quantum - as long as it is not measured or interacted with in any way - exists in a state in which all its possible real states are superimposed."*[70]

However, this does not mean denying the existence of matter and claiming that

---

[69] Fritjof Capra, "The Tao of Physics. An Exploration of the Parallels between Modern Physics and Eastern Mysticism", Flamingo, 1982, pp. 77-78.

[70] Ervin Laszlo, "Science and the Akashic Field. An Integral Theory of Everything", Inner Traditions, Vermont, 2007, p. 135.

everything is pure "spirit" or consciousness. This position has also been advocated in philosophy, but is of course highly questionable. If I walk through the flat and am not careful, I can bump my head on a door. Ouch! Now I've got a big bump. So matter is very real for us. If I cross the road and don't notice the bus approaching me at high speed, it will quickly convince me that it was made of more than just consciousness. So how can these two positions be reconciled? If matter is supposed to have emerged from a kind of "cosmic consciousness" in the sense of an energetic oscillation or vibration field, how can I explain the bump or the relentless collision with the bus?

The answer is that what we define as matter could simply be a ***lower-level energetic form*** of expression that emerged in the course of the cosmic evolutionary process. The combination of subatomic „particles" to form atoms, which then combine to form molecules, which in turn form more complex compounds and thus ultimately enable organic life, does not exclude this possibility. Understood in this way, what we generally consider to be "dead matter" - e.g. a stone or various metals - would merely be a stage in the evolutionary process that has not yet entered into higher-level, more complex, i.e. more living compounds up to organic life forms. To clarify my self-description, I would therefore be an ***evolutionary*** pan- or cosmopsychist:

> *"Evolutionary panpsychism neither reduces the whole of reality to structures consisting of intrinsically inert and insentient material building blocks (like materialism), nor does it equate the whole of reality with a qualitative non-material mind (like idealism).* ***It regards both matter and mind as fundamental elements of reality****, but (unlike dualism)* ***does not claim that they are radically separate;*** *they are* ***different aspects of the same reality****. What we call »matter« is the aspect we perceive when we look at a person, a plant or a molecule from the outside; »mind« is the aspect we get when we look at the same thing from the inside."*[71]

We can therefore also question the dualistic view, according to which a clear boundary could be drawn between "matter" and "spirit". This separation took place several centuries BC in the form of a paradigm dispute between the Greek philosophers Heraclitus (520 - 460 BC) and Parmenides (520/515 - 460/455 BC). The former assumed a world that is in a constant state of change, a world of continuous "becoming" that arises from a tension between opposites, which he in turn regarded as a unity: this unity transcends all opposites and was described by Heraclitus as "logos" (an idea that, as you have already learnt, reappeared many

---

[71] Ervin Laszlo, "Science and the Akashic Field. An Integral Theory of Everything", Inner Traditions, Vermont, 2007, p. 111.

centuries later, for example with the German philosopher Hegel). Parmenides, on the other hand, declared change to be an illusion of the human senses. According to him, the world is based on an indestructible, unchanging substance from which the various material/sensually perceptible phenomena emerge. This division gave rise to the tradition in the history of the humanities and sciences that assumes mind and matter to be completely different from each other. However, there have been attempts to overcome this divide and harmonise the two positions:

> *"In the fifth century BC, Greek philosophers attempted to overcome the sharp contrast between the views of Parmenides and Heraclitus. In order to harmonise the idea of unchanging being (of Parmenides) with that of eternal becoming (of Heraclitus), they assumed that being manifests itself in certain unchanging substances whose mixing and separation cause changes in the world. This led to the concept of the atom, the smallest indivisible unit of matter, which found its clearest expression in the philosophy of Leucippus and Democritus. The Greek atomists drew a clear line between spirit and matter and imagined matter as consisting of several "basic building blocks". These were purely passive and in themselves dead particles that moved in nothingness. The cause of their movement was not explained, but was often associated with external forces that were assumed to be of spiritual origin and fundamentally different from matter. In the centuries that followed, this image became an essential element of Western thought, of the dualism between spirit and matter, between body and soul."*[72]

After the Aristotelian world view had dominated thinking in philosophy and science for two thousand years - a predominantly spiritual world view that regarded the material level as unimportant, even negligible, as material phenomena were only seen as "shadows" of spiritual active principles, as more or less illusory and arising from human "sensory illusions" that were unable to penetrate to the spiritual core - it was not until the late 15th century that methods of investigation emerged that could be described as scientific in the modern sense - that is, in the sense of scientific rigour based on concrete observations, experiments and the subsequent evaluation of empirically collected data. It was not until the late 15th century that methods of investigation emerged that could be described as scientific in the modern sense - that is, in the sense of scientific rigour based on concrete observations, experiments and

---

[72] Fritjof Capra, "The Tao of Physics. An Exploration of the Parallels between Modern Physics and Eastern Mysticism", Flamingo, 1982, pp. 25-26.

the subsequent analysis of empirically collected data. Sophisticated mathematical models were developed to describe what was observed; often these mathematical models also led to speculations about a certain expected behaviour of a system, which was then either confirmed or refuted in the form of experiments.

> *"In classical physics, the mass of an object was always associated with an indestructible material substance, with a »stuff« of which all things seemed to be made. The theory of relativity showed that mass has nothing to do with a substance, but is a form of energy. However, energy is a dynamic quantity associated with activity or processes. The fact that the mass of a particle corresponds to a certain amount of energy means that the particle can no longer be considered as a static object, but as a **dynamic pattern**, as a **process involving the energy that manifests itself as the mass of the particle**."*[73]

Galileo Galilei (1564-1642) was the first scientist to combine empirical knowledge with mathematical precision - which is why he is considered the father of modern science. On a philosophical level, this was preceded by a development that led to an extreme mind/matter dualism - above all by the French philosopher René Descartes (1596-1650). "The "Cartesian" division", according to Fritjof Capra,

> *"enabled scientists to **treat matter as dead and completely separate from itself and to view the material world as a multitude of different objects assembled into a giant machine**. Isaac Newton advocated such a mechanistic view of the world, building his mechanics on this foundation and making it the basis of classical physics. From the second half of the seventeenth century to the end of the nineteenth century, the mechanistic Newtonian model of the universe dominated all scientific thought. Parallel to this, the image of a monarchical God emerged, who rules the world from above by imposing his divine law on it. The fundamental laws of nature that scientists were searching for were thus seen as unchanging and eternal laws of God to which the world was subject."*[74]

So can consciousness and thinking really only be traced back to material foundations

---

[73] Ibid., p. 88, my emphases.
[74] Ibid., p. 27, my emphasis.

- the lump of matter called the brain - or are there possibly other sources outside this reductionist framework? In his book "Shadows of the Mind", the British mathematician and theoretical physicist Roger Penrose posed a legitimate question:

> *"It seems difficult to maintain that mentality can be **entirely** separate from physicality. And if mentality is indeed connected with certain forms of physicality - and apparently **directly** - then the scientific laws that describe the behaviour of physical things so precisely must surely also say a lot about the world of mentality."*[75]

I would ask the following question: What if matter (physicality, in Penrose's words), which we call "dead", was merely a kind of preliminary stage in an evolutionary process that originally sprang from "mentality" (consciousness) and is structured or organised to realise itself through time in stages of organisation or complexity of increasing order? What if mentality and physicality - as they were understood in earlier stages of the development of scientific thought - were not to be regarded as separate levels, but as a holistic process? This question will not be discussed further at this point; this will only follow after a discussion of the quantum physical concept of non-locality.

To further explain this idea, I will use an example from the world of mathematics, which the British mathematician and theoretical physicist Roger Penrose provided in his book "The Road to Reality" using the so-called "Mandelbrot set". Some readers may have already heard of it. If not, I'm sure that most of you will have seen pictures of the so-called "apple men". Here are a few examples that I created using the software "Ultra Fractal 6.0". These structures are characterised by "self-repetition", by "iteration": in colloquial terms, this involves the constant repetition of either the same action or several similar actions (such as calculation instructions) in order to approach a certain solution (this idea will become important later when the question of the finiteness or infinity of the universe is raised again). Such "apple men" (and similar fractals) provide a very good and highly aesthetic visual representation of the problem mentioned above: How can human thought approach the infinite? Because no matter how often or how deeply you "zoom" into such a set - certain structures will repeat themselves infinitely.

---

[75] Roger Penrose, „Shadows of the Mind. A Search for the missing Science of Consciousness", Vintage Books, London, 1994, p. 203.

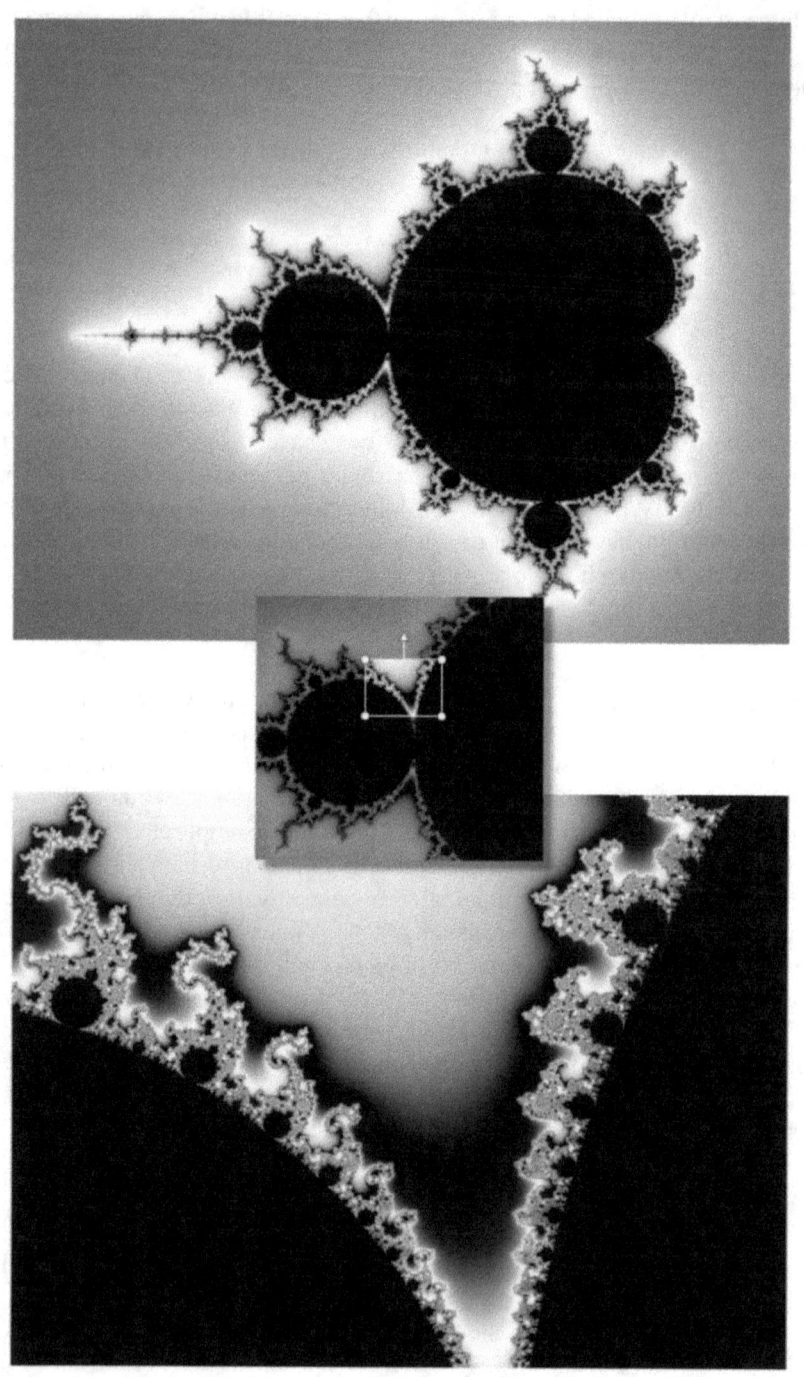

Four extremely strong enlargements from the edges of a so-called "Mandelbrot set". The principle of infinite iteration - and thus a sense of infinity - becomes clear in such images in a visually fascinating and highly aesthetic way. No matter how deep you look into any into any peripheral areas of the depicted structures - the "basic patterns" would would repeat themselves endlessly, so you would always encounter the same structures again and again.

Penrose raises an extremely interesting question in this context:

> "The point I want to make is that no one, not even Benoit Mandelbrot himself, when he first saw the incredible intricacies in the fine details of the set, had any real conception of the extraordinary richness of the set. The Mandelbrot set was certainly not an invention of a human mind. The set is simply objectively present in mathematics itself. If it makes sense to attribute an actual existence to the Mandelbrot set, then this existence does not lie in our minds, because no one can fully grasp the infinite variety and unlimited complexity of the set. Nor can it exist in the multitude of computer printouts that only begin to capture something of its incredible sophistication and detail, for these printouts give at **best a shadow of an approximation of the set itself**. Yet it has a robustness that is beyond doubt; for the same structure reveals itself - in all its perceptible detail, in ever greater subtlety the more closely it is scrutinised - independently of the mathematician or computer examining it. **Its existence can only lie within the Platonic world of mathematical forms.**"[76]

In another of his books - "Shadows of the Mind. A Search for the Missing Science of Consciousness" - Penrose also explores the question of why we humans are even able to grasp mathematical laws that sometimes lie outside our own thinking and imagination, and at least approximately express them.

> "You could perhaps describe the situation by saying that God is a very good mathematician and has used very advanced mathematics in building the universe. Our weak mathematical attempts allow us to understand the universe a little, and if we continue to develop higher mathematics, we can hope to understand the universe better."[77]

The most obvious answer would be: Since the universe and everything in it quite obviously obeys certain laws - what we generally refer to as the "laws of nature" - it would only be logical that our brains also developed according to such laws - which,

---

[76] Sir Roger Penrose, "The Road to Reality", Vintage Books, 2004, p. 16-17, my emphases.

[77] Paul Dirac, „The Evolution of the Physicist's Picture of Nature", Scientific American, Nr. 5, 1963

just as obviously, must have existed long before the arrival of humans on the stage of history. This is trivial, because we are "products" of this universe. Assuming that the assumption of materialistic science is correct, that our consciousness and thinking have no other cause or source than the material one in the form of the highly complex "circuits" in our brains and their electrical impulses, it would therefore only be logical that thinking that has developed in this way would be able to grasp or perhaps better: "depict" these laws on which it is based, because in this case our brains and the thinking that has emerged from them would themselves be an "image" of these laws of nature and thus of the mathematics that is reflected in them. To rephrase: In that case, our brains and our thinking would be „in-formed" expressions of an evolutionary process by which consciousness unfolds itself; by which a „non-local" field of energy develops into manifold „local" existence.

> *"What we determine mathematically is only to a small extent an objective fact, to a greater extent an overview of possibilities."*[78]

In other words, *our* mathematics would be a "humanised" or "condensed" image of what Penrose called the "Platonic world of mathematical forms" - the "world of the spirit" - reduced to the capacities of human thought. Or as one would say in the ancient mystery religions: a manifestation and spiritual reflection of "archetypes". In Plato's words: coming from the sphere of "pure forms".

> *"[Kurt] Gödel apparently took the view that the mind is not in fact constrained by the fact that it must be a computational entity, and that it is not even constrained by the finiteness of the brain. In fact, he scolded [Alan] Turing for not accepting this as a possibility. (...) Gödel did accept both of Turing's implicit claims that »the brain basically functions like a digital computer« and that »the laws of physics have a finite limit of precision in their observable consequences«, but rejected Turing's other claim that »there is no mind separate from matter«, labelling it »a prejudice of our time«. Gödel thus seems to have assumed that the physical brain itself must behave computationally, but that the mind is something that lies beyond the brain, so that the actions of the mind are not*

---

[78] Werner Heisenberg, "Schritte über Grenzen" (Steps over boundaries), Piper Verlag, München 1971, p. 90

> *compelled to behave according to the computational laws that he believed must govern the behaviour of the physical brain."*[79]

It would logically follow from this that it does not necessarily require **human** intelligent observation to make such laws "real" - i.e. that they are not based on mere human invention - because they have existed since the beginning of the cosmos anyway. Humans would therefore have merely "rediscovered" them through their own observation.

In quantum physical terms: In order to trigger the "collapse of the probability function", i.e. the transition from the state of mere probability or possibility - the merely probabilistic - into concrete, more precisely: localised material/substantial existence, the mere fact of the existence of such regularities could already suffice, **insofar** as one wishes to understand this as a kind of self-observation or self-measurement of a system - a self-development process - of which human thought would only be a partial expression or a „sub-function" that arose in the course of that biological process that we call evolution, through the fact that a sufficiently complex organ such as the human brain was developed that has the ability to "participate" in this process. Penrose:

> *"Somehow the very world of physical reality seems almost mysteriously to emerge out of Platonic world of mathematics. [...] Then there is the [...] mystery of how it is that perceiving beings can arise from out of the physical world. How is it that subtly organized material objects can mysteriously conjure up mental entities from out of its material substance?"*[80]

In my view, Penrose answered the first sentence himself: Why should it be a „mystery" that perceptual beings can emerge from a „physical world" which ***itself*** possibly emerged from a „Platonic world of mathematics" – in so far as you want to define this world as the world of consciousness? And why should it be mysterious that such beings do not conjure up „spiritual entities" out of their „material substance", but rather **embody** these spiritual entities in a physically „condensed" form as their living, dynamically (evolutionary) developing expression?

---

[79] Roger Penrose, „Shadows of the Mind. A Search for the missing Science of Consciousness", Vintage Books, London, 1994, p. 128
[80] Roger Penrose, „Shadows of the Mind. A Search for the missing Science of Consciousness", Vintage Books, London, 1994, p. 413-414

> "The best known of the interpretations that assume a fundamental role of thought and consciousness is attributed to a group of eminent physicists whose contributions were central to the development of quantum physics - John von Neumann, John Archibald Wheeler and Eugene Wigner; among them are a Nobel Prize and almost every other award ever given to physicists. Their joint interpretation is known as the von Neumann-Wheeler-Wigner interpretation (vNWWI). It is based on the premise that thought and consciousness are not just ephemeral artefacts of the brain's biochemical processes, **but that thought and consciousness are the fundamental causal elements of the cosmos.** You could say that vNWWI is the polar opposite of scientific materialism: it assumes not only that thought and consciousness exist beyond matter, **but that thought and consciousness create matter.**"[81]

The problem of consciousness was described very well from a quantum physics perspective in an article in the American scientific journal "Scientific American":

> "Quantum mechanics is science's most precise and powerful theory of reality. It has predicted countless experiments and produced countless applications. **The problem is that physicists and philosophers disagree about what it means, i.e. what it says about how the world works.** Many physicists - probably most - adhere to the Copenhagen interpretation, which was developed by the Danish physicist Niels Bohr. However, this is a kind of anti-interpretation, which says that physicists should not try to make sense of quantum mechanics; they should "shut up and do the maths", as the physicist David Mermin once put it.
>
> Philosopher Tim Maudlin laments this situation. In his book »Philosophy of Physics: Quantum Theory«, published in 2019, he points out that several interpretations of quantum mechanics describe in detail how the world works. These include the GRW model proposed by Ghirardi, Rimini and Weber, David Bohm's pilot-wave theory and Hugh Everett's many-worlds hypothesis. But here's the irony: Maudlin is so diligent in pointing out the flaws in these interpretations that

---

[81] Joseph Selbie, "The Physics of God", New Page Books, 2021, p. 125-126, my emphases.

he reinforces my scepticism. They all seem hopelessly half-baked and absurd.

Maudlin does not address interpretations that describe quantum mechanics as an **information theory.** *Positive perspectives on information-based interpretations can be found in »Beyond Weird« by journalist Philip Ball and »The Ascent of Information« by astrobiologist Caleb Scharf. In my opinion, however, information-based interpretations of quantum mechanics are even less plausible than the interpretations analysed by Maudlin.* **The concept of information makes no sense without conscious beings sending, receiving and reacting to the information.**

The introduction of consciousness into physics undermines its claim to objectivity. *Moreover,* **as far as we know***, consciousness only occurs in certain organisms that only existed on earth for a short time.* **So how can quantum mechanics, if it is a theory of information and not of matter and energy, apply to the entire cosmos since the Big Bang?** *Information-based theories of physics seem like a throwback to geocentrism, which assumed that the universe revolved around us. Again, given the problems with all interpretations of quantum mechanics, agnosticism seems to me to be a reasonable stance.*

[...]

I am definitely a sceptic. I doubt we will ever know if God exists, what quantum mechanics means and how matter creates mind. **These three mysteries, I suspect, are different aspects of a single, impenetrable mystery at the heart of things.** *But one of the joys of agnosticism - perhaps the greatest joy - is that I can keep searching for answers and hope that a revelation awaits me over the horizon."*[82]

The following statement from the excerpt is unfortunately not very satisfactory, I would even describe it as very debatable: *"**The introduction of consciousness into***

---

[82] John Horgan, „What God, Quantum Mechanics and Consciousness Have in Common", https://www.scientificamerican.com/article/what-god-quantum-mechanics-and-consciousness-have-in-common/, my emphases.

physics undermines its claim to objectivity".

Not at all. Why? Because, as the author of the article notes himself, according to *what we know,* consciousness should only occur in certain organisms, including us humans? But that's just a theory - not proof.

In other words, there is no evidence so far that creaturely consciousness - whether in any organisms as mentioned in the article or in us humans - is the ***only one*** that exists or that there is no other form outside of it. Why should consciousness be limited only to the human conceptions of it that have been developed so far? To repeat the question posed above: Can a separation between objectivity ("material world") and subjectivity ("human consciousness") - i.e. according to the dualistic model - be safely asserted and maintained at all? If one takes the pan- or cosmopsychistic view - as I do - then mind or consciousness itself ***would already be an objective reality***. The author writes that he doubts that "we will ever know [...] how matter generates mind". But perhaps it is the other way round: that mind/consciousness generated matter - if one imagines consciousness as a form of energy that underlies all existence in all its various manifestations.

„The concept of information makes no sense without conscious beings sending …". I agree. But what if „consciousness" (energy field, quantum vacuum, „Ain Soph", „God", take your pick) could be considered the ***sender*** of this primordial information that spawned material existence? What if we are the ***receivers***?

> *"Ultimately, matter is just a wave-like disturbance in the quasi-infinite sea of energy and in-formation that is the unifying field and the enduring memory of the universe."*[83]

Roger Penrose, whose example of the Mandelbrot set has already been cited, explains the problem as follows:

> *"I am aware that there will still be many readers who find it difficult to attribute any kind of actual existence to mathematical structures. Let me ask these readers to broaden their conception of what the term »existence« can mean to them. The mathematical forms in Plato's world clearly do not have the same kind of existence as ordinary physical objects such as tables and chairs.* ***They have neither a spatial location***

---

[83] Ervin Laszlo, "Science and the Akashic Field. An Integral Theory of Everything", Inner Traditions, Vermont, 2007, p. 105.

> *nor do they exist in time. Objective mathematical concepts must be regarded as timeless entities and **not as having been conjured into existence at the moment they are first perceived by man**. The particular vortices of the Mandelbrot set [...] did not come into existence at the moment they were first seen on a computer screen or a printout. Nor did they arise when the general idea behind the Mandelbrot set was first voiced by humans - not really first by Mandelbrot, as it turned out, but by R. Brooks and J.P. Matelski, in 1981 or perhaps earlier. For certainly neither Brooks nor Matelski, and initially not even Mandelbrot himself, had any real idea of the sophisticated detailed designs [...]. **These designs »existed« from the beginning of time, in the potentially timeless sense that they would necessarily reveal themselves in the exact form in which we perceive them today, no matter at what time or place a perceiving being wished to examine them.**"[84]

These basic mathematical patterns or "archetypal" forms could therefore exist in a form that does not correspond to our conventional notions of space and time - they could exist in a **non-local** form, as pure potentiality, as a "tendency" towards realisation in an existentially concretised, that is, **observable** sense.

> "The physicist Roger Penrose (1989, 1994) and the anaesthetist Stuart Hameroff (1998) have advocated a model according to which consciousness arises from quantum effects that occur in subcellular structures within neurons, the microtubules. The model assumes so-called "objective collapses" in which the quantum system moves from a superposition of several possible states to a single, unambiguous state without the intervention of an observer or measurement, as is the case in most quantum mechanical models. According to Penrose and Hameroff, the environment within the microtubules is particularly suitable for such objective collapses, and the resulting self-collapses generate a coherent flow that regulates neuronal activity and enables non-algorithmic mental processes."[85]

---

[84] Sir Roger Penrose, "The Road to Reality", Vintage Books, 2004, p. 17, my emphases.
[85] Robert van Gulick, „Consciousness", https://plato.stanford.edu/entries/consciousness/

At this point, I would like to briefly return to the question of the finiteness or infinity of the universe on the basis of what we have discussed so far. The following train of thought may seem too abstract, possibly even absurd, to some readers. However, I would ask you to approach it as impartially as possible and to put aside for a moment everything you have learnt from conventional wisdom. Above all, we are concerned here with the accuracy of conceptual thinking.

Pure potentiality - i.e. the mere probability of existence in the sense of the probability/wave function of quantum mechanics - is infinite by definition. This is because this potentiality not only describes **all** possible states of a system (and there are potentially an infinite number of ways to existentially concretise/localise a merely probable system), but it can also be realised (existentially concretised) **anywhere** and at **any time**. It is not yet concretely localised, e.g. in the form of a "real existing" space-time. Or a specific object in a space. To use a simple everyday example: In the form of a tennis ball lying on a table. In its material form - its existential concretisation as a material object that can be perceived by human observers - the tennis ball has finite dimensions, i.e. it has a certain circumference, a certain weight and is located at a definable point in a certain space (for example, on the table in my living room). As a "pure idea" - as a mere probability of a table tennis ball in the form of a "table tennis ball wave function" - it is **potentially everywhere at the same time**, as the idea or concept of a table tennis ball **has not yet experienced any spatio-temporal-existential "fixation"**. In its purely potential or ideal form, it is **non-local**, it does not yet have a time or a place - only as a substantially concretised table tennis ball does it become *local* - a concrete, observable thing. The table tennis ball could therefore "appear" wherever it is "observed" or where it is concretely "measured". To remind you once again: I am aware that such comparisons are of course a bit flawed.

> *"The basis of quantum physics is the assumption that the wave function contains **all possible information** about a system. The state of a system is completely defined by the wave function. But the wave function only has a probability interpretation."*[86]

Analogue to this: The mere possibility (probability) of space is per se infinite, since space can be created anywhere. See the quote above: The wave or probability function contains **all** possible information about a system. As long as there is no space at all, but it only exists as pure probability, there are logically an infinite

---

[86] C. Allan Boyles, "God and Quantum Physics", Wheatmark, 2021, p. 100

number of possible places - and points in time - to concretise/manifest it existentially - in the sense of spatiotemporal existence. The same applies to a "particle" in it.

> *"The probability wave introduced a completely new concept into theoretical physics. [...] It meant something like a **tendency towards a certain event**. It meant the quantitative version of the old concept of δύναμις or »potentia« in Aristotle's philosophy. It introduced a **peculiar kind of physical reality that stands roughly halfway between possibility and actuality**. [...] It was not a three-dimensional wave of the type of elastic or radio waves, but a wave in a multi-dimensional **configuration space**, which had only become known through Schrödinger's investigations, i.e. a rather abstract mathematical quantity."*[87]

See how quickly that happened? And here we are again in a dilemma of spatial and temporal („local") human thinking, its dimensional concepts. Any child would ask after the above: How can a space be created "everywhere" if none exists yet? At this point at the latest, human imagination reaches its limits again: How should a possible space be able to emerge "everywhere" from a merely potential space, which is supposed to be infinite due to its as yet unrealised spatiotemporal concretisation, if there is as yet no space *in* which this probability or wave function can be manifested as a concrete space? Surely the word "everywhere" per se presupposes the existence of a space in which something can arise? The same applies to time, of course: how can something be existentialised "at any time" if - according to human concepts - there is no time at all, but only - what is that supposed to be? - "potential" or "probable" time? Or, as Werner Heisenberg put it in the above quote: you have to imagine this "space" as a "multi-dimensional configuration space", whereby the terms "space" and "imagination" should be understood here in a "transempirical" sense, as they *exceed our possibilities of experience*.

> *"However, the waves associated with the particles are not »real« three-dimensional waves such as water or sound waves, but »probability waves«; abstract mathematical quantities*

---

[87] Werner Heisenberg, "Quantentheorie und Philosophie", Philipp Reclam Jun. Verlag GmbH, p.17-18, my emphases.

> *that refer to the probabilities of finding the particles in different places and with different properties."*[88]

In addition to the Big Bang theory, there have been other explanations around for a long time, such as that according to which space is eternal and infinite and filled with a certain form of energy, namely "quantum vacuum energy" (QVE), also known as "zero point energy" or "space energy". The American painter, sculptor, architect, philosopher and mystic Walter Russell (1871-1963) will tell us more about this "zero point" later.

> *"We are beginning to see the entire universe as a holographic, interconnected network of **energy and information** that is organic and **self-referential** at all levels of its existence. We and all things in the universe are **non-locally connected** to each other and to all other things in a way that is **not constrained by the previously known limits of space and time.**"*[89]

This brings us to a somewhat more detailed description of the concept of "non-locality". It originates from quantum physics and was developed to explain the phenomenon of "quantum entanglement". The term "quantum entanglement" was used to describe the strange phenomenon observed in experiments that information can be transmitted between two particles simultaneously, i.e. without any loss of time. What one particle "knows", its "sibling particle" also knows - without any loss of time. Depending on the distance between these particles, however, this would contradict Einstein's theory of relativity, according to which nothing can be faster than light:

> *"Ever since Einstein and his colleagues introduced the idea of quantum entanglement, physicists have been trying to solve the obvious violation of the speed limit of light. There were three possibilities. The experiments used to prove entanglement have significant flaws. But after many years and thousands of experiments, no significant flaws have been found. Secondly,*

---

[88] Fritjof Capra, "The Tao of Physics. An Exploration of the Parallels between Modern Physics and Eastern Mysticism", Flamingo, 1982, p. 166.

[89] Ervin Laszlo, Jude Currivan, „Cosmos: A Co-creator's Guide to the Whole World", Hay House, 2008, p. 8

information can travel through the universe faster than the speed of light. If this were the case, physicists would have to abandon the theory of relativity, which is based on the speed of light. Thirdly, the counter-intuitive concept that the physical universe is permeated by a non-local realm must be accepted.

Non-locality is a cumbersome term used by physicists to describe an area where there is no distance. *An object or event is considered local if it is subject to the effects of distance. Magnetic fields, for example, lose their strength over distance. Light takes time to travel from the sun to the earth. These are local effects. In contrast, objects or events in a non-local area are not affected by distance, which is counterintuitive.* **Since the world we perceive with our senses is always associated with distances, it is difficult for us to imagine such a realm.**

At first, non-locality was regarded as an abstract but meaningless artefact of the mathematics of quantum physics. But in 1964, John Stewart Bell put forward a theorem in an essay »On the Einstein-Podolsky-Rosen Paradox« that proved that non-locality can not only be a property of reality, but that it must be according to the mathematical foundations of quantum physics itself. In other words, you cannot have one without the other. If quantum physics is true - and it has proven this in countless applications - then non-locality is a real property of the cosmos."[90]

American quantum physicist and philosopher David Bohm (1917-1992) explored this idea of "non-locality" in detail:

"David Bohm did not trivialise the implications of "quantum weirdness", but embraced them. By embracing non-locality in particular, he was able to discover mathematically the order in the universe that had eluded Einstein. Bohm's work was groundbreaking. He developed a new mathematical system, now called Bohmian mechanics, and an interpretation of quantum physics bears his name. [...] Appearances aside, he discovered mathematically that nothing can be separate from anything else because the universe and everything in it is

---

[90] Joseph Selbie, "The Physics of God", New Page Books, 2021, pp. 75-76, my emphases.

> *invisibly connected to a two-dimensional, non-local realm. [...]* **He called this non-local area, which connects everything, pre-space. Pre-space is, as the name suggests, spaceless.** *In non-local pre-space there is no distance as we understand it. According to Bohm, entangled photon twins, when measured by an intelligent observer, instantly enter our space-filling three-dimensional world from this spaceless pre-space. Bohm's work shows that if entangled photons instantly appeared billions of light years apart in our physical universe, they would still have been in spaceless pre-space the moment before they appeared, and thus there would have been no distance between them at all the moment before they appeared."*[91]

And again we stumble across the inadequacies of man-made concepts: a space in which there are no distances at all would of course not be any space at all - not according to our conventional ideas. This is why quantum physicists rightly speak of "counterintuitive" concepts in such cases, because they contradict our usual experiences of spatiality. Perhaps it would therefore be better to speak of a "**pre-state**" instead of a "pre-space" in this case.

> *"The subatomic world is characterised by rhythm, movement and constant change. However, it is not random and chaotic, but follows very specific and clear patterns."*[92]

It was this concept of non-locality that inspired me (in connection with the idea of a "cosmic wave function" or probability function) to develop the ideas that I outlined above: A mere probability/potentiality of space can therefore naturally be everywhere at the same time (infinite), and if such a probability becomes existentially concrete - according to our human ideas of existence - this could mean that the existentially concretised space is also infinite, because it is a realisation or "concrete existential reflection" of an "infinite" (everywhere and simultaneously existing idea, an infinite "possibility of realisation"). Incidentally, this is precisely why the so-called "many worlds theory" was developed in quantum physics: an infinite potentiality could theoretically - as soon as it "collapses" into concrete

---

[91] Ibid., pp. 76-77, my emphasis.
[92] Fritjof Capra, "The Tao of Physics. An Exploration of the Parallels between Modern Physics and Eastern Mysticism", Flamingo, 1982, p. 273.

existence - also create an infinite number of conceivable universes.

> *"Quantum theory has shown that particles are not isolated grains of matter, but **probability patterns**, connections in an inseparable cosmic web. [...] The particles of the subatomic world are not only active in the sense that they move very quickly; they are themselves processes! The existence of matter and its activity cannot be separated from each other. They are just different aspects of the same space-time reality."*[93]

That's what I meant by the transition from non-locality to locality. Just as in Bohm's work, entangled photons that appear very far apart in our universe, e.g. millions of light years apart, can exist in a "pre-space" (or better, pre-state) before they appear, in which there is no distance between them. In this sense, it would not even be necessary to *exchange* information between them - because they come from the same "pre-state" and therefore already carry the same information when "entering" a three-dimensional space, i.e. four-dimensional space-time - even if there are huge distances between them in the local space - the one that we can perceive. Analogous to the example of the 2D world inhabited by two-dimensional beings, the possibly gigantic distance that lies between these "sibling particles" could therefore possibly only be what we see when a higher-dimensional structure "crosses" a four-dimensional spacetime or "spans" it like a "cosmic umbrella".

At this point you will certainly remember the cosmogonic ideas of mystical-religious coinage, for example the Kabbalistic concept of the "Ain", from which the "Ain Soph" emerges through a "concentration of God on himself" - a primordial existence, the "I am".

I will take the liberty of repeating some of the ideas presented so far, this time, however, commenting in square brackets on quantum physical terms or concepts that you have now become familiar with. It goes without saying that these comments are not of a scientific, but of a speculative philosophical nature.

Firstly, the Kabbalah:

- "AIN SOPH was described by the Kabbalists as the oldest of all the ancients. [...] They symbolise the essence of AIN SOPH by a circle, which in turn is a symbol for eternity **[the infinite possibilities of existential concretisation; the wave or probability function of the infinite "spirit";**

---
[93] Ibid., p. 225, my emphasis.

this function describes all conceivable/possible states of a system; as long as there is no existential manifestation in the sense of observability, there are only - as Hans-Peter Dürr, for example, put it - "possibilities of a realisation", a "tendency" to existence; space and time exist in this "pre-state" only as pure potentiality that can be realised anywhere and at any time]. This hypothetical circle encloses a dimensionless realm of incomprehensible life [non-locality, therefore incomprehensible, unimaginable for human, dimension-bound thinking and imagination], and the circular boundary of this life is abstract and measureless infinity. According to this concept, God is not only a centre, but also a realm. Centralisation is the first step towards limitation. [Collapse of the probability function ("spirit"), i.e. the mere possibility/probability of existence into an existential concretisation; transition from infinity to finiteness, i.e. to "limitation"]."

- "Substances, essences and intelligences are manifested out of the inscrutability of AIN SOPH, but the Absolute itself is without substance, essence or intelligence." [See above: all existing things are "localised" manifestations out of a non-locality, a dimensionless, infinite state of mere potentiality. This of course has no substance or intelligence, it is mere "can-be-ness" of unfolding, evolutionary causes – including the development of beings capable of intelligent thought]

The position of Friedrich Wilhelm Joseph Schelling:

- "The first, a reflective movement, is the attempt of the infinite to embody itself in the finite." [The "infinite" - the mere potentiality, the probabilities of the wave function as a description of all possible states - requires an "opposition", a counterpart, in order to be able to perceive itself as "something" at all, in this case as a mere possibility or "spirit". However, this view logically presupposes that the tendency or impulse to create a counterpart must already be inherent in this "potentia"; it remains questionable whether this must necessarily be a conscious act or an extremely abstract logical necessity "of itself" that is incomprehensible to human comprehension; perhaps human ideas of consciousness are completely inadequate here]

The hermetic teachings:

- "In this way it was accomplished, O Hermes: the Word, moving like a breath through space, called forth the fire by the friction of its movement. [...] This will continue from an infinite beginning to an infinite end, for the beginning and the end are in the same place and in the same state." **[The "fire" - the spirit/the potentiality/non-locality that is absolutely at rest in itself - set itself in "motion", i.e. it began to vibrate (a "word was spoken") - whereby spatio-temporally concretised (localised/observable) existence came into being: because every vibration, however small it may be, naturally needs space in order to be able to vibrate, and also time, because without the passing of time nothing can vibrate/not move. Existence emerged from motionless non-locality, which is why its beginning and end are "in the same place and in the same state" ("pre-space" or "pre-state")]**

Roger Penrose:

- "The mathematical forms in Plato's world clearly do not have the same kind of existence as ordinary physical objects such as tables and chairs. They have neither a spatial location nor do they exist in time." **[They originate from the abstract realm of non-locality, inaccessible to human conceptualisation, in which these forms exist only in probabilistic form, as probabilities of existence - in the form of the probability function. Only through the "collapse" of this probability function, i.e. after the transition from the "realm of possibility" to that of existential concretisation, do they take on what we call "real forms", which we can examine and thereby determine (observe) certain regularities].**

American author Dr Rhawn Gabriel Joseph expressed the idea of a self-conscious creative energy in his article "The Quantum Physics of God: How Consciousness Became the Universe and Created Itself" as follows:

> *"If consciousness is energy, then the energy that is the quantum continuum is also likely to become conscious. If the universe as a whole is a manifestation of the quantum continuum as perceived by consciousness, then the continuum could have become conscious of itself and, in achieving self-consciousness, created the universe, which created itself by becoming conscious."*[94]

As Joseph also deals with controversial topics that are generally regarded as "crackpot", he is accused of practising pseudoscience. In any case, the above quote contains completely legitimate questions that other scientists - including numerous Nobel Prize winners in physics, who certainly cannot be suspected of working in a

---

[94] Dr. Rhawn Gabriel Joseph, „Quantum Physics of God: How Consciousness Became the Universe and Created Itself",
https://www.researchgate.net/publication/344906681_Quantum_Physics_of_God_How_Consciousness_Became_the_Universe_and_Created_Itself

dubious manner - have already raised before him. This should be clear from the quotes collected so far.

> *"It is important to realise that in today's physics we do not know what energy is. We have no idea that energy occurs in small lumps of defined size."*[95]

What speaks against this idea? If the human brain can develop consciousness and thinking due to the enormous complexity of its "wiring" - its neuronal network - and assuming that this complexity is an essential cause or source, why should the idea that a quantum continuum of far greater complexity of "wiring" (quantum entanglement) could also produce consciousness and intelligence be "absurd" or "ridiculous"? Just because we cannot prove it with the methods we have developed so far? Just because this intelligence possibly surpasses human intelligence to an extent that we cannot determine with our methods of measuring IQ? And because this intelligence possibly "communicates" in completely different ways that elude our sensory perception or the capacity of our thinking and language?

In order to stay a little longer with the problem of a sufficient definition of "consciousness" in the context of the discussions here, the following is an excerpt from an article in the "Rudolphina" of the University of Vienna:

> *"Quantum physics challenges our idea of what reality is. [...] Caslav Brukner is a quantum fundamentals researcher:* **he is looking for laws that remain valid even when time and observation can no longer be defined** *[...] "Physics is the attempt to describe the world in its real modes of action, independently of our observations. But quantum physics seems to be shattering this dream," says the Professor of Quantum Fundamentals and Quantum Information Theory. This is because the individual measurement result cannot be predicted at the smallest scale and observations can only be explained using probabilities: "The decay of atoms, for example, can only be described in the language of quantum physics, which is probabilistic - in other words, we always talk about probabilities," explains Brukner. [...] Classical physics describes the world deterministically - that is, as predictions*

---

[95] Richard P. Feynman, „Vorlesungen über Physik" (Lectures on Physics), vol. I, chapter 4.1 (Translation: Heinz Köhler), p. 46, Oldenbourg, Munich, Vienna, 5. Ed. 2007

of measurement results that are certain to occur under ideal conditions. Probabilistic statements, on the other hand, can only indicate a probability for a certain measurement result. There are also probabilistic statements in classical physics, "but these are only an expression of our ignorance of the true circumstances," says Caslav Brukner. "However, this is not the case in quantum physics, where probabilities are fundamental and non-reducible - there is no deterministic cause behind them."[96]

I object: If time and observation can no longer be defined, laws would be meaningless. In such a state, they would exist at most as mere possibilities, as probabilities of (natural) laws and thus also as mere probabilities of observability - they would therefore be in a kind of "mental limbo", which - according to quantum physics - only creates something existentially concrete, measurable and definable through observation. Regularities can only be discovered through observation; they describe observable processes. Since nothing can take place without the passage of time, in such a state laws would either be unnecessary or at most serve a self-sufficient existence in the form of mere probability. One could also say that they would be "frozen" in this state as a mere abstract possibility/probability.

To reiterate, "mind" - defined here as an abstract "collection" or "cloud" of possibilities or probabilities of existence (which only exists in a "probability space") - could not know what it is if there is still nothing - really nothing else besides it - through which such self-perception and self-definition („self-observation") could succeed. In order to make this possible, something must be created, quasi as a "point of reference". If a "mind" or a "consciousness" wants to know that it is one, an intentionality must be present: The "desire for knowledge", in this case perhaps better: the desire for self-awareness or to become ever more self-aware in an evolutionary process of unfoldment.

"The self can be seen as the point of perspective from which the world of objects can be experienced (Wittgenstein 1921). It offers not only a spatial and temporal perspective for our experience of the world, but also one of meaning and comprehensibility. ***The intentional coherence of the realm of experience is based on the double interdependence between the self and the world:*** *the self as a perspective from which*

---

[96] Daniel Schenz, „Es gibt keine Realität jenseits der Beobachtung", 1/23/2023, https://rudolphina.univie.ac.at/quantenphysik-es-gibt-keine-realitaet-jenseits-der-beobachtung, my emphasis.

> objects are recognised, and the world as an integrated structure of objects and events whose experiential possibilities **implicitly determine the nature and place of the self** (Kant 1787, Husserl 1929).'[97]

However, **human notions** of intentionality would probably be out of place: not in the sense of conscious human action, but probably more in the form of a logically necessary, abstract **information-processing process.**

> 'Ultimately, it will be (at least) as difficult to strictly define a mental state as a material state.'[98]

To illustrate how complicated this discussion, which I am breaking down here into a few terms and simple analogies, can very quickly become, I would like to quote from an article that deals with this question in great detail. You have already come across some of the terms: Physicalism, panpsychism, information theory as well as some terms from quantum physics. As a layman, you can hardly get through it. We are talking about a large number of individual fields of research, which are often difficult enough in themselves - and often disagree with each other in their results.

> "The problem becomes even more acute when you turn to other species. For centuries, research into consciousness in nature has been hampered by a firm belief in human exceptionalism. But the diversity and complexity of animal behaviour has laid this belief to rest, at least among biologists. This is particularly true for mammals. In psychophysical tasks involving the pressing of simple buttons, trained macaque monkeys behave very similarly to human subjects, signalling even when they see nothing. Visual self-perception, metacognition (knowledge of one's own mind), theory of mind, empathy and long-term planning have been demonstrated in primates, rodents and other orders. [...] When we consider species that are increasingly distant from Homo sapiens in evolutionary and neuronal terms, it becomes more difficult to establish

---

[97] Robert Van Gulick, „Consciousness", https://plato.stanford.edu/entries/consciousness/, my emphases.
[98] Harald Atmanspacher, Edward N. Zalta, „Quantum Approaches to Consciousness", https://plato.stanford.edu/entries/qt-consciousness/

> *consciousness. Two observations, one related to the complexity of behaviour and the other to the complexity of the underlying nervous system, are crucial. First, ravens, crows, magpies, parrots and other birds, tuna, coelacanths and other fish, squid and other cephalopods, bees and other members of the vast class of insects are all capable of sophisticated, learned, non-stereotypical behaviours that we associate with consciousness when performed by humans [...] Second, the nervous systems of these species exhibit an enormous and as yet misunderstood complexity. The bee contains about 800,000 nerve cells whose morphological and electrical heterogeneity is comparable to that of a neocortical neuron. [...] On the other hand, the lessons we have learned from the study of behaviour (...) and the neuronal correlates of consciousness in humans must make us cautious when it comes to inferring the presence of consciousness in creatures that are very different from us - regardless of how sophisticated their behaviour and how complicated their brains are. [...] The extent to which non-mammalian species share with us the gift of subjective experience thus remains elusive."* [99]

Many different branches of science are working together on an interdisciplinary basis to create a "theory of consciousness". This naturally includes neuroscience and brain research, behavioural psychology, cognitive theory and psychology and, as already mentioned, even quantum physics, especially the relatively new research field of quantum biology and information theory.

But what should information theory have to do with this? The reason is that information is exchanged between a certain "system" such as an organism and its environment (i.e. another "system"). Which raises the question: **what kind** of information from the environment affects an organic, living system, possibly influencing it to make a leap in its development, and through **which** "channels" is this information transmitted or exchanged?

> *"Despite the obvious advantage of simple life forms capable of rapid replication, living systems have reached different levels of cognitive complexity in terms of their potential to cope with environmental uncertainty. Given the unavoidable costs associated with recognising and adaptively responding to environmental cues, we hypothesise that the potential to*

---

[99] Giulio Tononi, Christof Koch, „Consciousness: here, there and everywhere?", https://royalsocietypublishing.org/doi/10.1098/rstb.2014.0167

> *predict the environment may overcome the costs associated with maintaining costly, complex structures. We present a minimal formal model, based on information theory and selection, in which successive generations of agents are modelled as senders and receivers of a coded message. Our agents are guessing machines, and their ability to deal with environments of varying complexity defines the conditions for obtaining more complex agents."*[100]

For the sake of simplicity, the authors of the article assume that machines are learning systems. They themselves point out that the processes taking place in organic organisms can be far more complex - which is why their approach should be seen more as an attempt to approximate the problem.

The next question we could ask is: How does an organism in turn affect its environment? The latter question is of course of great interest, especially in connection with the "Copenhagen interpretation" or the "collapse of the probability function" through human observation/measurement. Please do not be put off by the abstract language of the technical article:

> *"Any discussion of state collapse or state reduction (e.g. by measurement) refers, at least implicitly, to superposition states, since these are the states that are reduced. Since entangled systems remain in a quantum superposition [superposition state, author's note] as long as no measurement has taken place, entanglement is always included when talking about state reduction. In contrast, some of the dual quantum approaches use the topic of entanglement differently and independently of state reduction. Inspired by and analogous to entanglement-induced **non-local** correlations in quantum physics, mind-matter entanglement is conceptualised as the hypothetical origin of mind-matter correlations. **This presents the highly speculative picture of a fundamentally holistic, psychophysically neutral level of reality from which correlated mental and material realms emerge.**"*[101]

To put it more simply: there were already research approaches that assumed a "holistic", form of reality on a "probability level" in a natural philosophical-

---

[100] Luís F. Seoane, Ricard V. Solé, „Information theory, predictability and the emergence of complex life", https://royalsocietypublishing.org/doi/10.1098/rsos.172221
[101] Harald Atmanspacher, Edward N. Zalta, "Quantum Approaches to Consciousness", https://plato.stanford.edu/entries/qt-consciousness/

speculative manner and with the inclusion of quantum physical findings, from which spiritual and material areas emerge, between which there is a connection (correlation). In such explanations, "spirit" and "matter" are therefore not treated as strictly separate from each other, not as two different levels.

Coming back to consciousness, the overlaps between information theory and panpsychism are particularly interesting:

> "*IIT* [integrated information theory of consciousness, author's note] *subscribes to the panpsychist intuition that consciousness may be present throughout the animal kingdom and even beyond, albeit to varying degrees. All else being equal, integrated information, and thus the richness of experience, is likely to increase with the number of neurons and the richness of their connections, although the sheer number of neurons is no guarantee, as the cerebellum shows.*"[102]

The complexity of information-processing structures on the material level alone, which includes the human brain, is therefore no guarantee of the existence or development of consciousness.

> "*Primas (2003, 2009, 2017) proposed a dual approach in which the distinction between mental and material domains results from the distinction between two different modes of time: tensed (mental) time, including unawareness, on the one hand, and tenseless (physical) time, considered as an external parameter, on the other. [...] Considering these two concepts of time as implied by a **symmetry breaking of a timeless, psychophysically neutral level of reality**, Primas presents the tensed time of the mental realm as quantum-correlated with the parameter time of physics via »time entanglement«.*"[103]

This "symmetry break" - one could also speak of an *imbalance* that emerges from a "timeless, psychophysically neutral" level - will become somewhat clearer in the context of Walter Russell's ideas about the actual nature of the universe.

---

[102] Cit.op.
[103] Harald Atmanspacher, Edward N. Zalta, "Quantum Approaches to Consciousness", https://plato.stanford.edu/entries/qt-consciousness/, my emphases.

> *"In the same way, the IDEA that appears in matter is not in matter. The IDEA is never created. IDEA is a quality of the spirit. The IDEA never leaves the omniscient light of consciousness. The idea is only imitated by matter in motion. [...] All knowledge, all energy and all methods of creating any thing are qualities of consciousness alone. In matter, which is movement, there is no knowledge, no energy, no life, no truth, no intelligence, no materiality and no thinking."*[104]

To repeat it one last time to conclude my train of thought: In my opinion, the "paradox of the intelligent observer" is a supposed one **because it did not consider the possibility of the existence of a non-human intelligent observer/observation as the root cause for cosmic existence** (as far as I understand it).

The aforementioned Dr Rhawn Gabriel Joseph explained further in his article:

> *"Consciousness exists. Consciousness has energy. Energy can become matter and matter can become energy. Consciousness is always a consciousness of something.* **Consciousness needs a duality in order to exist as consciousness.** *Self-consciousness, the consciousness of consciousness, is also a duality. As Heisenberg, one of the founders of quantum mechanics, summarised: "The transition from the possible to the actual takes place in the act of observation ... and the interaction of the object with the measuring device and thus with the rest of the world ... Since observation has caused our knowledge of the system to change discontinuously, its mathematical representation has also undergone the discontinuous change, and we speak of a quantum leap" (Heisenberg, 1958). In other words: something comes into being by becoming aware of it.*
>
> *[...]*
>
> *The act of recognising, observing or measuring, i.e.*

---

[104] Walter Rusell, „Eine neue Vorstellung vom Universum" (A new concept of the Universe), Genius-Verlag 2019, p. 29-30

> *interacting with the environment in some way, creates an entangled state and a node in the quantum continuum that is described as a "collapse of the wave function"; an energy node that represents a kind of flaw in the continuum of the quantum field. This quantum node concentrates at the point of observation, at the assigned measured value, and can be entangled. Consciousness perceives a flaw in the continuum, but it is still part of the continuum, even if it is perceived as separate.* **The universe exists because there is a consciousness of the universe.**
>
> *[...]*
>
> *As Neils Bohr and Werner Heisenberg, the founders of quantum theory, emphasised, there are direct parallels between quantum mechanics and Taoism, Buddhism and Hinduism: "The great scientific contributions to theoretical physics ... have ... a relationship between the philosophical ideas in the tradition of the Far East and the philosophical substance of quantum theory."*[105]

If you want to call this consciousness "God" or "Creator", it does not automatically follow that this is the same as the "created". A carpenter is not the chair he has made or the sculpture he has carved from a tree stump - but he needs these things to be able to define himself as a carpenter. A chef is not the exquisite five-star menu that she has prepared for her guests. A composer is not the piano sonata he wrote - but he absolutely needs it to be able to identify himself as a composer. An automotive engineer is not the vehicle she designed - but she absolutely needs it to be able to define herself as an automotive engineer. Defining yourself as an author without ever writing would be pointless.

In the same sense, a "God" - a creative, intelligent consciousness (in "non-localised" form, to use the terms of quantum physics for a while) - could require concretised existence in order to "assure" itself of its creative potentialities (in a concretely existentialised, i.e. "localised" form); as already mentioned, this could therefore be an absolutely necessary basic condition for conscious existential self-perception or observation to be possible at all.

---

[105] Dr. Rhawn Gabriel Joseph, „Quantum Physics of God: How Consciousness Became the Universe and Created Itself",
https://www.researchgate.net/publication/344906681_Quantum_Physics_of_God_How_Consciousness_Became_the_Universe_and_Created_Itself, my emphases.

This brings us to Walter Russell, briefly mentioned above, who in his book "A New Concept of the Universe", published in 1953, on whose German translation [106] I participated, is based on precisely this idea: Similar to the "vNWWI" (von Neumann-Wheeler-Wigner interpretation), matter is not the causal factor, but thought and consciousness are the causal, fundamental elements of the universe.

> *"Energy is indeed the substance from which all elementary particles, all atoms and therefore all things in general are made, and at the same time energy is also what makes things move."*[107]

Russell speaks of a "universal zero point of rest", one could also say: absolute spiritual (energetic) immobility, from which an imbalance arises through conscious ("creative") thinking, which is essentially based on polarity - on an interdependence of two directions of movement. According to Russell, this involves "condensation" or "focussing" towards a point (this idea is reminiscent of the Qabbalah's creation narrative of the "Ain" and "Ain Soph"), which he describes as the cause of gravity, as well as the opposite movement, "relaxation" - what we call "radiation" (radioactivity), i.e. the dissolution or "unwinding" of "matter" into space. According to Russell, space and matter are inextricably linked.

> *"The one outstanding, superordinate property of our electric universe of opposing, balanced movement effects is the cyclical unfolding of matured body forms into the manifestation of the idea of consciousness and their refolding into the primal source of all idea."*[108]

This state of absolute motionlessness is the "pivotal point" of an equilibrium from which the two opposing directions of movement of mental "concentration" and "relaxation" create the universe or matter in motion like a lever, i.e. an imbalance.

> *"Energy belongs to the invisible universe. It extends ONLY THROUGH A pivot point, which is at rest, into the visible universe of motion.* **However, energy does not extend beyond**

---

[106] Walter Russell, „Eine neue Vorstellung vom Universum", Genius Verlag, Bremen, 2019
[107] Werner Heisenberg, "Physik und Philosophie" (Physics and Philosophy), 7. Ed., Stuttgart: Hirzel, 2006, p. 92
[108] Walter Russell, „Eine neue Vorstellung vom Universum", Genius Verlag Bremen, 2019, p. 144

> *the pivot point into matter, into the state of matter or into the motion of matter. **That which extends beyond the state of rest into motion is an expression of energy - an imitation of energy - an effect that emanates from a cause to show what energy is capable of doing when projected into the illusion of motion.***"[109]

Russell sees the cardinal error of the natural sciences, especially physics, in viewing matter as the cause or causal factor and searching for the "ultimate reason" of all things in it - for him, matter is merely an *effect*, namely an "effect in motion". This is created by the "creative spirit" setting itself in motion - for him it is a matter of conscious thinking - i.e. beginning to vibrate (from the zero or "pivotal point" of its universal rest) - and thereby creating what we call "material reality". Russell therefore regards this realm of the "spirit-idea" as the actual reality (in which he agrees with the cosmogonies of Hinduism, Buddhism and Taoism as well as Kabbalah). He also calls it the "light of the cause". He also calls the two opposite directions of movement "states of pressure", which, depending on the degree of their compression, give the impression of different substances, e.g. the elements of the periodic table. For him, however, these are "tones" of different "octaves". For example, there would be the hydrogen octave or the carbon octave.

> *"The undivided and unrestricted light of consciousness is an eternal state of rest. This invisible light of the spirit is the equilibrium of perfect balance and absolute motionlessness, which forms the basis of the universe of movement divided into states of pressure. [...] **The light of cause, divided into the two opposite qualities of light of effect**, is the only activity of consciousness that we call thinking."*[110]

According to Russell, electricity is the only force in the universe that does work:

> *"Electricity is an effect of effort, tension and resistance. These are produced in the light of the mind by the energy of the desire to divide and expand the balanced unity of the one resting light of universal consciousness into pairs of many polar divided units of thinking consciousness. [...] Electricity thus does the »work« of the world by striving for separateness and multiplication of units, and also by relaxing from resistance to such stresses and tensions until the vibrational movements cease and retreat into the universal state of rest. The only*

---

[109] Ibid., p. 57., my emphasis.
[110] Walter Russell, „Eine neue Vorstellung vom Universum", Genius Verlag, Bremen, 2019, p. 35-37

>>work<< done in this universe is that >>work<< which is caused by the exertion and tension of electrically divided matter in motion."[111]

Please allow me once again to compare Russell's ideas with the cosmogony of the Kabbalah: In the latter, there was talk of a "vacuum of pure spirit", the "Ain", from which, through a "concentration of the spirit on itself" ("self-observation"; "collapse of the probability function"; transition from the non-locality of the purely probabilistic into observable existential concretisation/localisation), the "Ain Soph" emerged - a primordial existence). The completely motionless state of the spirit in absolute stillness thus set itself in motion - began to oscillate, to vibrate - and thereby created a polarity, a relationship of tension between unequal states.

> "The vacuum turns out to be a cosmic medium that transports both photon waves (light) and density pressure waves, exerts the force that can ultimately decide the fate of the universe, and gives mass to the particles we know as »matter«. Such a medium is not an abstract theoretical entity. It is not a vacuum, but a physically real and active plenum."[112]

As silly as it may sound to some people, let's nevertheless drop all inhibitions for a moment, forget everything that is considered scientifically serious and "approved" today, everything we have learned at school and university, everything we know from the press, radio and television - and speculate: It is possible that electricity or electrical energy is a force that arose from a - in this case quite literal - tension in that an extremely abstract "pre-state" (the "pre-space" of a David Bohm), unthinkable for us humans, was looking for a way to "experience" itself in the form of a consciousness, i.e. to be able to begin a process of becoming self-aware. To this end, a bipolar relationship was initially created: Non-existence/existence, probability/concretisation, mere possibility of reality/observable, measurable reality. Maybe this is the source of the energy we call „electricity".

> "As long as they are not observed or measured, quanta have no unique properties, but exist simultaneously in several states. These states are not »real«, but »virtual« - they are the states that the quanta can assume when they are observed or

---

[111] Ibid., S. 42-43
[112] Ervin Laszlo, Jude Currivan, „Cosmos: A Co-creator's Guide to the Whole World", Hay House, 2008, p. 66.

> *measured. It is as if the observer or the measuring device fishes the quanta out of a sea of possibilities."*[113]

In modern physics, there is the well-known concept of matter and antimatter. Concrete example: For every electron there is a positron, i.e. an anti-particle of the same mass and opposite charge. There is therefore a polarity, a relationship of interdependency. Now let's interpret this a little more boldly and generously and put forward the thesis that the so-called "quantum field theory" could be used to support Russell's views. The first essential aspect of quantum field theory is

> *"... the fact that all interactions involve the creation and destruction of particles, such as the absorption and emission of the photon* [by an electron, author's note]; *and the second feature is a **fundamental symmetry between particle and antiparticle**. For every particle there is an antiparticle with the same mass and opposite charge. The antiparticle of the electron, for example, is called the positron and is usually labelled e+. The photon, which has no charge, **is its own antiparticle**. Electron and positron pairs can be **spontaneously generated by photons** and transformed into photons by the reverse process of annihilation."*[114]

Pairs of electrons and positrons can therefore be generated spontaneously **by photons** - with a little imagination, one could draw a parallel here to Russell's "resting" or "undivided, unrestricted light of consciousness", which ensures a "universe in motion" by generating pairs of opposites (polarity). This is because the photon is "its own antiparticle" - it can therefore produce both particles, both matter and antimatter. However, there is an important distinction to be made: According to Russell, there is a difference between the light that is visible to us and the invisible, "magnetic light of God", as he calls it. This idea in turn is reminiscent of the Kabbalah, in which a primordial point of existence ("Ain Soph") emerges from the "Ain", the "pure vacuum of the spirit" ("Potentia", the mere possibilities of the probability function) through a kind of concentration towards a centre (self-observation/measurement), from which in turn the "boundless light" ("Ain Soph Aur") arises.

---

[113] Ervin Laszlo, "Science and the Akashic Field. An Integral Theory of Everything", Inner Traditions, Vermont, 2007, p. 26.
[114] Fritjof Capra, "The Tao of Physics. An Exploration of the Parallels between Modern Physics and Eastern Mysticism", Flamingo, 1982, p. 199-200, my emphases.

Russell also claims that there are not just two, but four magnetic poles:

> "*His* [Johannes Kepler, 1571 - 1630, German astronomer, astrologer, physicist, mathematician and natural philosopher, author's note] *law of elliptical orbits proves that he was on the verge of discovering that four - not two - magnetic poles control the twofold opposite equilibrium of this counter-rotating universe. With only two magnetic poles, a three-dimensional, radial universe of time intervals and cyclic sequences would be impossible. A universe in equilibrium must have two poles to control centripetal, genero-active force and two balancing poles to control centrifugal, radioactive force.*"[115]

In Russell's work, we often find the same ideas as in Hinduism, for example: the universe of "matter" is only an "illusionary" one; in reality, it is a matter of vibrational fields that merely create the impression of absolutely solid substance (one could call them "standing waves"). According to Russell, the natural sciences need to rethink their basic assumptions based on the senses:

> "*The senses have not shown man that our universe is only an immaterial universe of movement. Nor have they made him aware of the principle of polarity, which divides the universal equilibrium into pairs of oppositely constituted partners in order to create a sexually divided, electric, counter-rotating universe.*"[116]

If Russell were alive today and someone were to explain to him how scientists at the CERN particle accelerator are trying to search for the "primordial ground" of existence in matter, he would probably sigh: "Oh boy, they're still looking in the wrong place":

> "*It would be just as reasonable to scratch to the bottom of Leonardo da Vinci's painting* The Last Supper *to find a fundamental particle of colour and its first brushstroke, in the hope of finding the IDEA of the painting and also its creator. [...] Science is still searching for the original principle of life in matter as eagerly as it searched for the original substance from which all other substances arise. [...] The time has come in the spiritual unfoldment of man when he must recognise that*

---

[115] Ibid., p. 12-13
[116] Ibid., p. 13

> *all IDEA lies eternally in the zero-point equilibrium of the resting magnetic light of the cosmic spirit - which is God - and that IDEA manifests only through the movements of body forms in polarised circuits. These appear from the eternal zero point and must disappear again into this zero point so that they can return in endless cycles."*[117]

It is perfectly clear that such statements would of course immediately be categorised as "controversial", especially nowadays. More than that: many would probably even label Russell a total "nutcase" because he talks about "God". What is equally certain - and I can confirm this from my own experience of debating with such people - is that most of those who are so incensed by this and who rub this word up the wrong way cannot begin to fill it with meaning. Instead, they rant about how it is still possible to use such anachronistic words in an "enlightened" age characterised by scientific and technological rationalism. God, you gotta be kidding me. This is the Middle Ages, alas, the Stone Age!

If you do get answers, these answers are often quite depressing: The institutionalised church (the Vatican) and its indeed rather ungodly, bloody past are brought up, buzzwords such as "inquisition" are thrown at you, you are asked whether you still believe in UFOs, the "flat earth" or that God kneaded people out of clay and then breathed life into them, or the "nonsense about Adam and Eve in paradise".

But that's the way it is in an increasingly mindless and thoughtless age, in which there is apparently hardly any understanding for **allegories and parables**; in which unfortunately only a few people ask the question of whether all this can and should be understood literally at all, or whether such stories are perhaps to be understood more metaphorically/allegorically. **This is precisely the case.** The Hindu idea of the dancing god "Shiva", for example, does not of course mean that a supernatural being on LSD would have danced his name as the universe - in truth, as Fritjof Capra correctly noted in his book "The Tao of Physics", this cosmogony contains profound insights in metaphorical form that can actually be harmonised with those of modern physics. This is why the quote from Nobel Prize winner in physics Max Born should be repeated here:

> *"We have reached the end of our journey into the depths of matter. We have searched for solid ground and found none. The deeper we penetrate, the more restless the universe becomes, the more vague and cloudy. [...] There is no fixed place in the universe: everything rushes and **vibrates in a wild***

---

[117] Ibid, p. 88-89

dance."[118]

So when I repeatedly write "God" in this book, I am by no means a fanatic from the "Bible Belt" or a religious zealot who wants to convert his fellow human beings to the "right faith". Rather, I am simply asking whether this very nebulous, diffuse, vague word "God" could possibly be used to describe something that human thought is unfortunately unable to define more precisely due to its limitations. It is, so to speak, only a "placeholder", a linguistic symbol for such complex causal relationships and processes, for such an overwhelming vitality of diverse modes of expression (existential manifestations in space and time), that the attempt to narrow them down linguistically would amount to fixing a butterfly the size of the sun with pins in a thimble.

Furthermore, it cannot be implicitly assumed that the beginning of the cosmos must have been overly complex. After all, it is possible that a fairly simple, manageable set of initial conditions can lead to increasing complexity over time. It is therefore not possible to draw a simple reverse conclusion: The mere fact that the universe, our planet, its biological diversity and our own life and thinking appear to us to be overwhelmingly complex does not prove that the cause of this existence must also have been equally complex or even more complex:

> *"A finite and **surprisingly simple set of basic elements controlled by a small set of algorithms** can generate great and seemingly incomprehensible complexity, if only the process **is allowed to unfold over time**. A set of rules that informs a series of elements **sets in motion a process that orders and organises the elements so that they form ever more complex structures and relationships."*[119]

The American physicist James Hartle (1939 - 2023) proposed a similar model - based on the quantum physical idea of the "superposition" (superimposition) of many possible realizable (existentially concretized) states in the wave or probability function, i.e. in the state of mere potentiality:

> *"But although the dynamical laws of classical physics are simple, the universe itself is complex - and so its initial state must have been complex too. Describing the exact positions*

---

[118] Max Born, Nobel laureate in physics, from the postscript of his book "The Restless Universe", Springer-Verlag, 1969, p. 166, my emphasis.

[119] Ervin Laszlo, "Science and the Akashic Field. An Integral Theory of Everything", Inner Traditions, Vermont, 2007, p. 12, my emphases.

and momenta of all the particles involved requires so much information that any statement about the initial state is too complex to be a law. Hartle suggested that quantum mechanics can solve this complexity problem. **Since the wave function of a quantum object is distributed over many "classical" states [...], one could propose a simple initial condition that contains all complexities as emergent structures in the quantum superposition of these states.** *All observed complexities can* **be seen** *as* **partial descriptions of a simple fundamental reality: the wave function of the universe.** *As an analogy, a perfect sphere can be decomposed into many parts with complicated shapes, but which can be reassembled into a simple sphere."*[120]

In other words, this "wave function of the universe" could thus be described as a "folded-in universe" in a non-local state, which only creates an evolutionarily developing (existentially "localized") sphere that unfolds through time through observation/measurement. Whereby, of course, the question arises again as to whether this "collapse" of the "probability wave" from the state of the non-local - the potential/probabilistic - of the universe into a locally realized space-time was actually just coincidence or a conscious act.

Sometimes talk of a "higher reason" and the idea of an intelligent order behind things can even trigger very aggressive reactions - especially when someone makes the "mistake" of using the word "God", which is frowned upon for some reason. Like the widely known American physicist and string theorist Michio Kaku, for example. One small example is enough to illustrate a problem that I pointed out right at the beginning of this book: Sticking too closely to man-made concepts that would first have to be defined in more detail.

In an interview, Kaku once said that he suspected a kind of organising reason, an intelligence behind the universe. A completely legitimate statement, one might think, which also coincides with the beliefs of numerous high-ranking scientists who have already had their say in this book. So what's the problem? Nevertheless, he was quickly accused of lacking seriousness - he had made himself "untrustworthy".

"For some reason, the compatibility of science and religion is constantly and loudly proclaimed in respectable intellectual circles," it said in an article on the website "Why Evolution is true".

---

[120] Eddy Keming Chen, „Does quantum theory imply the entire Universe is preordained?", https://www.nature.com/articles/d41586-023-04024-z, my emphases.

> *"I don't know exactly why that is - we'll have another example tomorrow from Smithsonian Magazine, of all places - but here we see the well-known popular scientist and theoretical physicist Michio Kaku promoting God in an article in Intellectual Takeout in June: »World-famous scientist: God created the universe«. Kaku, who specialises in string theory, is a professor of physics at the City University of New York. And the title clearly plays into the hands of those of an Abrahamic persuasion. But seriously? God created the universe? Well, let's ask Dr Kaku for his evidence - »evidence« he originally provided in CNS News. And here, as far as I can see, is the entirety of his evidence: There are laws of physics."*[121]

Why so offensive right away? Firstly, why should it be problematic that the title of an article "plays into the hands" of those who have an "Abrahamic mindset"? Are people of this faith now under general suspicion? Have they committed a crime or "sinned" against the sciences? Secondly: Before you get hung up on a word like "God", you first have to fill it with meaning. **Concrete** meaning. After all, it's just a three-letter word: G,o,d. It is a collection of the most diverse ideas about an ominous entity that is supposed to have created something. It's just a belief. I've **tried** to give it a *possible* meaning in this book - that's it. No need to get excited. The vast majority of heated debates simply stem from serious misunderstandings and, above all, very imprecise, laboured terminology.

Kaku himself responded to the accusation as follows:

> *"There's a website that misquoted me. That's one of the disadvantages of being in the public eye: sometimes you get misquoted. And the reference I saw said that I had said that you could prove the existence of God. My point of view is different. My own position is that you can neither prove nor disprove the existence of God."*[122]

Elsewhere he is said to have said:

> *"I have come to the conclusion that we are in a world that*

---

[121] „Michio Kaku embarrasses himself, says that the laws of physics and the behavior of subatomic particles reflect »the mind of God«", https://whyevolutionistrue.com/2016/11/17/michio-kaku-embarrasses-himself-says-that-the-laws-of-physics-and-the-behavior-of-subatomic-particles-reflect-the-mind-of-god/

[122] „Michio Kaku Clears Up God Discovery", https://innotechtoday.com/michio-kaku-clears-god-discovery/,

> *functions according to rules created by an intelligence [...] Believe me, everything we call chance today will no longer make sense. [...] It is clear to me that we exist in a plan that is governed by rules created and moulded by a universal intelligence and not by chance."*[123]

Kaku is not alone in this opinion - if quoted correctly - as can be seen from the quotes of numerous scientists collected in this book. If one wants to give this "universal intelligence" - and again one could draw parallels, for example, to the "Ain" of the Kabbalah ("vacuum of pure spirit") or the Indian "Brahman", the "world soul" - the name "God", this is not a scientific question, but first of all a natural or existential philosophical-speculative one, secondly one of personal convictions and thirdly, above all, one of language, of definitions of terms. This is why there is no reason for unproductive, superfluous bickering between philosophy and science: as long as it is not clear what is actually meant by all these terms - be it a vague "universal intelligence", a pan- or cosmopsychistic "consciousness" that is supposed to underlie everything, the Kabbalistic "vacuum of pure spirit" or "God" for all I care - personal attacks, invective and mockery are not only out of place, but also spiritually unfruitful.

To conclude the notes on Walter Russell's cosmogony, we will briefly discuss his ideas on "polarity", which according to him was a basic prerequisite for the creation of the universe:

> *"The two opposing states of pressure which govern the cycles of life and death of all bodies are:*
>
> *(a) the negative state of expansion, which strives radially and outward in spiraling paths from a central zero point of rest to form the state of low potential that creates »space« and*
>
> *b) the positive state of condensation that strives inward to a central zero point of rest to form the condensed state of gravity that creates bodies forming into solids surrounded by space.*
>
> *The desire of consciousness expresses itself through the electrical process of thought. Thought divides the idea into oppositely constituted units of motion that record a replica of the idea in thought form.*

---

[123] „World-renowned professor of theoretical physics: »We are in a world ... created by an intelligence«", https://www.thecollegefix.com/world-renowned-professor-theoretical-physics-world-created-intelligence/

> *Sir James Jean suggested that it might be possible to prove that matter is »pure thought«.* **Matter is not pure thought, but it is the electrical record of thought.** *Every electric wave is a recording instrument, eternally recording* **the forms of thought in wave fields of matter.**
>
> *All thought waves generated in any wave field* **become universal by being repeated everywhere.**"[124]

The polarity, according to Russell,

> "*creates moving body forms in pairs of opposites and places the counterparts of each pair on the opposite side of a common equator. In this way, both partners are so dependent on each other that they can only exist in constant exchange through interaction. [...] Each sun has its equal and opposite partner in the form of a black vacuum hole on the other side of its equator. [...] Gradually, the empty black hole becomes the sun and the sun becomes an empty black hole.*"[125]

To briefly summarize the most important points: For Walter Russell, "God" is the moving thing, the "driving force" behind cosmic existence, but he is not contained in the moving thing. For him, the phenomenal world we perceive is a material simulation of spirit-idea in motion. One could draw a comparison here to Plato's idea of the sphere of pure forms or the ancient Greek "archetypes", but Russell's world view is much more dynamic and alive: there are no "solid" particles in the sense of early atomistic theories, but only "idea" in motion, with electricity playing a central role: "Matter does not consist of pure thought, but it is the electrical record of thought."

It is remarkable and fascinating that the quantum physicist and philosopher David Bohm, also mentioned in this book, who appeared when the concept of "non-locality" was introduced, developed a very similar concept of the universe. Bohm also assumes an indivisible or unbroken "wholeness" of cosmic existence, thus also pursuing a holistic approach (non-locality was discovered as part of the famous "Einstein-Podolsky-Rosen experiment"). Bohm believes that order is inherent in the cosmos, i.e. that there is an implicit or inherent order which, according to him, could be found on a "non-manifest" level, i.e. either in the non-local or perhaps even in an area that has not yet been discovered. The fascinating thing is that Bohm speaks of

---

[124] Walter Russell, „Eine neue Vorstellung vom Universum", Genius Verlag, Bremen, p. 39, my translation and emphases.
[125] Ibid., p.108-109

an "enfolded" order - Russell, in turn, writes in his book "A New Concept of the Universe" of the "enfolded" spirit idea, which begins to vibrate from the "universal zero point of rest" by dividing into polarities. He uses the metaphor of a seesaw to create pairs of opposing movements ("particles"/"antiparticles"?) in order to existentially manifest the forms emerging from the "idea" in moving matter through "unfolding".

Equally interesting is Bohm's use of the hologram concept to support his idea: Just as in the S-matrix and "bootstrap" theory, which you learned about very early on in this book, the notion of "particles" was more or less abandoned in favor of talking about a tightly interwoven "web" or "mesh" of energetic processes and that each "particle" that is "detected" through observation/measurement potentially bears or mirrors all others within it and all others in turn bear or mirror this one. So a hologram, as we know, contains all the image information in each of its parts. If a hologram were to be broken up into small pieces, the entire image would still be revealed when one of the fragments is irradiated with light:

> *"Bohm uses the hologram as an analogy for this implicit order because of its property that each of its parts contains, in a sense, the whole. If any part of a hologram is illuminated, the whole image is reconstructed, even if it has less detail than the image obtained from the complete hologram. In Bohm's view, the real world is constructed according to the same general principles, with the whole contained in each of its parts."*[126]

This is of course also interesting because light also plays a fundamental role in Russell's cosmogony: he speaks of the "magnetic light of God" with which "he" controls, i.e. orders, his universe - again, a parallel could be drawn between Russell's idea and Bohm's idea of an "inherent" or "enfolded" order on a non-local level. To emphasize the processual, dynamic character of his concept, Bohm uses the term "holomovement".

> *"In order to understand the implicate order, Bohm found it necessary to consider consciousness as an essential feature of holomovement and to explicitly include it in his theory. He sees mind and matter as interdependent and correlated, but not causally connected. They are mutually enveloping projections of a higher reality that is neither matter nor consciousness."*[127]

---

[126] Fritjof Capra, "The Tao of Physics. An Exploration of the Parallels between Modern Physics and Eastern Mysticism", Flamingo, 1982, p. 352.
[127] Ibid., p. 353

I would once again like to express my own view that consciousness is something other than ***self-awareness***. I would like to emphasise the processual nature of our "reality": In contrast to earlier scientific world views - quantum physics has clearly shown this much - we must no longer assume "static", fixed objects or existential manifestations, but rather a dynamic "overall process" called the universe. In this I overlap with the British mathematician and philosopher Alfred North Whitehead.[128] I have drawn on the cosmogony of the Kabbalah for this: From the "Ain", the "vacuum of pure spirit" (quantum vacuum/quantum vacuum energy), a "collapse of the probability function" is triggered by a "concentration on oneself" (self-observation/definition), which creates existential manifestations in localised form (in space and time). From the Bohmian "implicit" or "folded-in" order, which exists outside the dimensions we can perceive, which therefore has "no time and no place" - which lies in the non-local, in the state of mere potentiality/probabilistic) and can therefore potentially be everywhere and simultaneously - a self-development process emerges, which could be what is summarised under the term evolution: The existential "testing" of these potentialities (possibilities of manifestation) in a space-time continuum whose "blueprint", however, remains in the "hologram" of the "pure spirit" and - in Russell's sense - is "projected" from this "universal zero point of rest" (the non-local) in the form of a universe in motion.

To conclude this chapter, a few comments on the question of what advanced scientific explanations there are for the highly complex phenomenon of human consciousness. It is now recognised that not only the genetic material DNA, but also entire cells actually emit light, so-called "biophotons".

> *"In recent years, it has become increasingly clear that photons play an important role in the basic functioning of cells. Most of this evidence comes from switching off the light and counting the number of photons that cells produce. To the great surprise of many people, it turns out that many cells, perhaps even most cells, emit light as they work. In fact, it seems that many cells use light to communicate. There is evidence that bacteria, plants and even kidney cells communicate in this way. Various groups have even shown that the brains of rats literally glow thanks to the photons generated by the neurons as they work. [...] And that raises an interesting question: What role does light play in the work of*

---

[128] Alfred North Whithead, „Process and Reality", Free Press, 1979.

neurons?"[129]

In quantum biology, it is assumed that cells possibly organise themselves through intercellular biophotonic communication, i.e. that they join together via such communication channels to form organs and ultimately the entire organism - be it a plant, an animal or a human being. What's more, it is also assumed that there could be an exchange of information at a **non-local** level throughout the entire organism. In concrete terms, this means that there are theories according to which biological organisms could be quantum systems on a macro level. The exchange and organisation of information transported by biophotons could provide an explanation for human consciousness.

> *"In recent years, Russian and German researchers (Popp, Voeikov and others) have made great progress in studying the biophysical aspects of biophotonic processes in humans. The present work suggests that the way in **which biophysical light interacts with human self-organisation of information**, which can be achieved through biomolecular, metabolic or neuronal communication, is a multi-part reality. These systems may merge as mobile energy relay systems, similar to what is seen in acupuncture science as Qi processes, suggesting a »holo-movement« that seeks to validate itself and increasingly retrieves and utilises only the information that serves its exchange with the environment. This co-evolution of evolutionary process levels, expressed in process terms, can be seen as the basis for a medicine of light that integrates hidden variables in consciousness research with functional differentiation and new insights in the biological sciences."*[130]

Now that some of the most important terms and concepts of quantum physics have been explained, let us conclude with another attempt to overcome the classical mind-matter dualism with regard to a possible explanation of the phenomenon called "mind" or consciousness. The focus - as I have made sufficiently clear - is on a „cosmopsychic" perspective.

Since quantum theory is the most comprehensive explanation of the behaviour of

---

[129] „The Puzzling Role Of Biophotons In The Brain",
https://www.technologyreview.com/2010/12/17/198375/the-puzzling-role-of-biophotons-in-the-brain/
[130] Bruce D. Curtis, J.J. Hurtak, „Consciousness and Quantum Information Processing: Uncovering the Foundation for a Medicine of Light",
https://www.liebertpub.com/doi/pdfplus/10.1089/107555304322848931, my emphases.

matter at the atomic and subatomic level to date, the first question to ask is, of course, to what extent quantum physical processes and effects - i.e. those at the microphysical level - should affect macrophysical structures and systems such as the human brain. I base all further explanations on Roger Penrose's book "Shadows of the Mind", as it is simply one of the best books on this subject and deals with the problems I am concerned with in the most differentiated and comprehensive way.

Penrose writes,

> "... that the implications of quantum theory must be relevant to brain activity, either by providing a role for quantum uncertainties or for non-local collective quantum effects (such as the phenomenon known as »Bose-Einstein condensation«)."[131]

According to Penrose, one of the main problems of quantum physics today is that certain levels of our physical reality cannot be explained precisely or without contradiction. The existing theory must therefore be expanded and modified in order to eliminate the most obvious paradoxes. The famous "observer paradox" has already been mentioned and I explained why it is possibly only a supposed one: because the "state vector reduction", i.e. the "collapse of the probability function" was, in my view, the cause of the creation of the universe: from a non-locality into an existentially concretised locality. As I have already said, I am not a scientist, but merely a scientifically literate philosopher. I am therefore fully aware that my assumptions may be flawed or even wrong. My theory is therefore only a very modest attempt to provide an explanation for the phenomenon called consciousness in the hope that it may provide some inspiration for scientists to develop my ideas further (if they prove plausible and scientifically viable).

Penrose is of the opinion that

> "... the pure randomness of existing measurement theory must be replaced by something else, in which essentially non-calculable components will play a fundamental role."[132]

Why these "components" might not be calculable because they are to be found in non-locality or in the non-local, I will explain further below. As I have already explained in sufficient detail in this book, I am of the opinion that there need be no contradiction here because the existing measurement theory (i.e. the theory of the factor of observation, which plays a central role in quantum physics) ignores the

---

[131] Roger Penrose, „Shadows of the Mind. A Search for the missing Science of Consciousness", Vintage Books, London, 1994, p. 204
[132] Ibid., p. 205

possibility that our universe could have emerged from such a "measurement"; that the collapse of the probability function - a process that Penrose calls "**R**" in his book - has long since taken place and that we, as *human* observers, have emerged from such a "total cosmic evolutionary process". This would solve the "paradox" of the observer to the extent that *our* human observation could only be a sub-process of a larger, more comprehensive process of "self-observation" and development of a system (in this case the universe).

To rephrase this once again: Penrose thus argues that there is something in human thought that goes beyond the mathematical models we have known so far; something that may not be "mathematised" at all, that is, that cannot be adequately described on the basis of the mathematical systems we have known and developed so far. He also writes in "Shadows of the Mind" that this may also involve quantum physical elements such as *non-local effects*. But what if this problem were only an apparent one?

In what way? Well, if our brains were a *localised* "expression" of a cosmic consciousness "collapsed" into existential concretisation through self-observation/measurement, emerging from a spatio-temporally unfolded evolutionary, i.e. process of *becoming „self-aware"*; if this consciousness ("quantum vacuum", the "vacuum of pure spirit" of the Kabbalah) exists in a non-local form and under the assumption that our own consciousness also originates at least in part from this non-local source or is at least connected to it in some way, for example through the phenomenon of quantum entanglement, i.e. that there are - as Penrose also assumes - non-local elements, which need to be taken into account but have not yet been sufficiently investigated and which run on our "localised", i.e. material brain ("hardware") as a suitable basis for "complex computing operations" ("software") - then, if my theory is correct, it would also explain why there is, as Penrose suggests, a place in our consciousness and thinking "that we cannot scratch". In other words, something that goes beyond "computability", as he calls it, i.e. beyond predictability or calculability.

> *"All these many different substances in this universe are merely many different states of pressure. These have been produced by the interaction of the opposite motion between two opposite poles of rest, which emerge from the zero-point universe of the **knowing mind** to imitate the manifold ideas of the **thinking mind.**"*[133]

After all, computability is only possible in the local: Non-locality is characterised by a state of pure probability, i.e. it exists in a form (without space and time) in which there are no observable/measurable processes that could be described by humans in the form of physical laws. Perhaps this is also the source of the mathematical "intuition" of which Penrose speaks; perhaps this is the "Platonic sphere of pure forms" or "archetypes".

Penrose:

> *"I am dealing here with a completely different set of issues, namely the question of what can in principle be perceived with the help of the human mind, thought and insight. It turns out that these questions are very subtle indeed, even if their subtlety is not immediately obvious. At first glance, these questions seem trivial, because correct thinking is surely just correct thinking - something more or less obvious, which in any case was already clarified by Aristotle 2300 years ago (or at least by the mathematical logician George Boole in 1854, etc.)! But it turns out that "correct thinking" is something very subtle and, as Gödel (with Turing) has shown**, lies beyond any purely computational activity.** These questions have historically been the domain of mathematicians rather than psychologists, and the subtleties involved have generally not been the concern of psychologists. We have seen, however, that these are questions that give us insight into the ultimate physical actions that must underlie the processes that underlie our conscious understanding. [...] These issues also touch on deep questions of mathematical philosophy.* ***Does***

---

[133] Walter Russell, „Eine neue Vorstellung vom Universum", Genius Verlag, Bremen, p. 92, my translation.

mathematical understanding represent a kind of contact with a pre-existent Platonic mathematical reality that has a timeless actuality independent of us; or do we create all mathematical concepts independently as we think through our logical arguments?"[134]

I suspect that the answer is: Yes, mathematical thinking represents a contact with a "pre-existent" reality - which Penrose calls "platonic-mathematical" - namely the **non-local superposition** of **all conceivable states** or potential possibilities of realisation, of which our universe represents **a** well-defined, existentially concretised form of expression. It is indeed timeless and independent of us insofar as we, as an evolutionary product of this well-defined form of existence in time, are only **one** possible expression of it, while in the non-local far more - in principle infinitely many - possible forms exist simultaneously - since these have not yet "collapsed" into a well-defined form and thus **remain in a state of pure potentiality**.

I am considering the possibility that what Penrose calls "mathematical intuition" - what he calls the "non-computable" factor, which gives humans the ability to "sense", i.e. intuitively grasp, mathematical laws even before they have been formulated in detail - is based on this non-locality. This could also be a plausible explanation as to **why**, as Penrose argues in "Shadows of the Mind", this non-calculable factor in human thinking exists at all: Because calculations logically always presuppose temporal processes. A calculation needs, trivially enough, time in order to be carried out. In the non-local, however, space and time only exist as a possibility of **realisation**, i.e. as potentiality/probability - as I understand it. What we define as calculation could therefore only exist in an already existing space-time, i.e. in localised existence - e.g. our existence.

However, this theory raises the question of why we should then be able to draw correct conclusions about the mathematical laws of our cosmos in this way - through the contact with the non-local in our consciousness, which we cannot fully grasp mentally, but which we can "intuitively" sense or sense - since there is, in principle, an infinite number of potentially realisable laws in the non-local. This, in turn, could be due to the fact that this intuitively perceived information originating from the non-local transforms ("collapses") in our brains into the well-defined form that defines the entire structure of our universe and thus, of course, us. In other words: in the transition from the non-local to the local - our conscious, observing, measuring thinking - this information "swings" to the "frequency" of our existence; the basically infinite possibilities are thus reduced to our form of existence. Perhaps it would be

---

[134] Penrose, „Shadows of the Mind", p. 208, my emphases.

helpful to let Penrose himself have his say on this:

> *"A quantum measurement has the effect that quantum events are magnified from the quantum level to the classical level. At the quantum level, linear superpositions [superposition, author's note] persist under the constant action of **U-evolution**. However, once effects are magnified to the classical level, where they can be perceived as **actual** events, we **no longer find things in these strange, complexly weighted combinations**. [...] This "jumping" of the state description of the system from the superimposed quantum state to a description in which one or the other alternative of the classical level takes place is called **state vector reduction** or **collapse of the wave function** (...)."*[135]

Oh dear. Another new term: "**U-evolution**". What is that again?

> *"U is described by the so-called Schrödinger equation, which specifies the rate of change of the quantum state or wave function in relation to time."*[136]

Penrose continues:

> *"The point is that the physical state $|\psi\rangle$ (determined by the ray of complex multiples of $|\psi\rangle$) is DETERMINED by the fact that the result YES for this state is CERTAIN. No other physical state has this property. For any other state, there would only be some probability, but no certainty, that the outcome will be YES, and an outcome of NO could occur. So, although there is no measurement that tells us what $|\psi\rangle$ actually IS, the physical state $|\psi\rangle$ is uniquely determined by what it states to be the result of a measurement that COULD be performed on it. This is again a question of counterfactuality [...], but we have seen how important counterfactual questions are for the expectations of quantum theory."*[137]

---

[135] Ibid., p. 263, my emphases.
[136] Ibid., p. 259
[137] Penrose, „Shadows of the Mind", p. 314-315.

Once again, we encounter the problem that I mentioned above: it is tacitly assumed that there is "no measurement" that "tells us" what the physical state expressed by the mathematical symbol $|\psi\rangle$ should be. In the case of our universe, there are well-defined, mathematically describable "rules" or laws of nature, which, according to quantum physics, presuppose measurement/observation. This observation/measurement or rather "self-measurement" of a state of infinite potentiality, i.e. in principle an infinite number of possible states, can be the cause or source of our physically and mathematically well-defined and describable universe. The contradiction - the "counterfactual" - only arises if this possibility is completely ruled out and not even considered. However, there is *no plausible, logically comprehensible reason* for this. If one assumes - as I do from my perspective as a cosmopsychist - that the "cosmic wave function" or - in the sense just explained - the "cosmic state vector" **could already be understood as consciousness or a state of consciousness** and triggers the reduction of infinite potentiality (the totality of all possible probable alternative states) through a "self-measurement" in order to create a very specific, well-defined state (our universe), there would be no contradiction.

"Although it may well be," continues Penrose,

> *"that the problem of mind is ultimately related to that of quantum measurement - or the U/R paradox of quantum mechanics - it is my belief that it is not consciousness per se (or consciousness as we know it) that can solve the internal physical problems of quantum theory. I believe that the problem of quantum measurement should be addressed and solved before we can expect to make any real progress on the question of consciousness in relation to physical action - and that the measurement problem must be solved in purely **physical** terms. [...] I believe that solving the quantum measurement problem is a **prerequisite** for understanding the mind, but **in no** way believe that it is the same problem. The problem of the mind is a much more difficult problem than the measurement problem!"*[138]

Consciousness could well solve the "internal physical problems of quantum theory" - provided it is not regarded as a **consequence** or **epiphenomenon** of physical processes, but as their **cause**. I fully agree with Penrose: solving the quantum measurement problem, i.e. the alleged "paradox of the intelligent observer", is a

---

[138] Ibid., p. 330-331.

prerequisite for understanding the phenomenon of consciousness/mind - however, I believe that physics has so far "put the cart before the horse", i.e. started in the wrong place by trying to maintain an artificial dividing line in the form of "mind-matter dualism". As I have already mentioned, for me as a cosmopsychist, consciousness is the fundamental level of what we regard as reality, or in the words of Max Planck:

> *"I regard consciousness as fundamental. I consider **matter to be derived from** consciousness. We cannot get beyond consciousness. Everything we talk about, **everything we consider to exist, presupposes consciousness.**"*

Penrose himself points out that it cannot be consciousness in the form "that *we* know" that can solve the existing problems of quantum theory. Now one can hold the view that only physical explanations could be satisfactory for the "measurement" or "observation" problem, or the one I am advocating here: that what *we* call consciousness and conscious thought *is* a physical phenomenon - an ***objective level of reality*** - which expresses itself in the form of human consciousness and thought as an individualised "partial aspect" of this reality. From this perspective, the human brain could represent a kind of "interface" through which this physical phenomenon expresses itself in and through us - even if we have not yet recognised this and therefore have so many problems with recognising consciousness as an objective level, ***because we only experience it in the form of our subjective, individual partial aspect in the form of our own consciousness in "ego form"***. Penrose:

> *"If the "mind" is something that lies outside the physical body, it is difficult to understand why so many of its attributes can be very closely associated with the properties of a physical brain".*[139]

Here I have to disagree for the following reasons: If one assumes that consciousness or "mind" is a fundamental level of physical reality (even if we have not yet recognised or acknowledged it as such) and considers the possibility, which I have just explained, that our universe emerged from this very level and that our brains are therefore also a product of this cosmic evolutionary development process, there would be no reason to believe that consciousness is a fundamental level of physical reality. If we assume that consciousness or "mind" is a fundamental level of physical reality (even if we have not yet recognised or acknowledged it as such) and take into account the possibility that I have just explained - that our universe emerged from this very level and therefore our brains are also a product of this cosmic, evolutionary unfolding process, there is certainly the possibility that mind is something that our brain has developed and "uses" as a suitable structure for its expression in individualised form, but that it can nevertheless lie "outside" the physical body in that it is not part of *local* physical existence, but is possibly to be found on another level of reality. But which one? Well, in the ***non-local***. Penrose writes that quantum theory is undoubtedly part of our physical reality - this cannot be denied. Quantum physics is therefore not a pure invention of human thought. Even if - and I agree with him here too - it needs to be developed further and requires some changes in order to resolve the contradictions inherent in it. The phenomenon of non-locality is therefore also part of this reality.

In this respect, it is worth asking whether, apart from pure conjecture, there is at least one or other scientific basis for making the kind of connections between locality and non-locality that I have just described. This does indeed seem to be a possibility:

> *"Of particular note in this context are recent observations of 35-75 Hz oscillations that appear to occur in association with brain regions involved in conscious attention. **These appear to have some puzzling non-local properties.**"*[140]

So the question is whether the non-local possibly contributes to our consciousness and if so, in what way and to what extent.

> *"Einstein was convinced that God does not play dice, that reality is ordered and determined. He hoped to find the laws for such deterministic behaviour within the **local** physical universe - but quantum mechanics ruled out this possibility. Bohm's deeper explorations of the strange side of quantum physics and the adoption of his holographic principle by string theory suggest that the order Einstein sought not only exists, but is also **non-local**."*[141]

Now one could - without wanting to split hairs - draw a distinction between consciousness and conscious thinking: Consciousness could be defined as a ***state***, whereas conscious thought could be defined as a ***process***. In quantum physical terms, and even if this is a very bold assumption, one could define the non-local superposition of possible unambiguous, well-defined physical states - which

---

[140] Roger Penrose, „Shadows of the Mind. A Search for the missing Science of Consciousness", Vintage Books, London, 1994, p. 374, my emphasis.
[141] Joseph Selbie, "The Physics of God", New Page Books, 2021, p. 81, my emphasis.

therefore exist in a "superimposed" state, and simultaneously, as claimed in quantum physics - as consciousness, in which no "state vector reduction", i.e. no "collapse of the probability function" has yet taken place. Only after this reduction - the unfolding into a spatio-temporal, localised process - can conscious thinking emerge *as a process*. Since both levels - locality and non-locality - are regarded as common elements of physical reality in quantum physics, there would therefore be no major contradiction. In this sense, "mind" would be a fundamental level of reality, as would its unfolding through state vector reduction into an evolutionary, spatiotemporal process.

Now that I have tried to underpin my existential-philosophical speculations at least a little with the findings of contemporary quantum physics, I will briefly return to the question posed at the beginning, whether existence could be a "logically necessary consequence" of non-existence, or in quantum-physical terms: Whether the mere probability - the possibility - of "locality", i.e. an existential concretisation through "state vector reduction" (collapse of the probability function) could have brought about a process of realisation out of non-locality. I had asked what "sense" it would make to exist in a merely probabilistic form, i.e. to "be there" only as a probability, without it ever coming to an observable existence. In other words, whether a "superposition" of possible/conceivable possibilities of realisation in the non-local already carries its existential concretisation in the form of a possible localisation to enable self-observation/measurement. Think back to the Kabbalah: The "vacuum of the pure spirit" created a point of existence called "I am" through a "concentration" of this spirit "on itself".

Were these all just completely unfounded speculations? Only absurd, crazy fantasies, which also don't have the slightest bit to do with "reality"?

Believe it or not, if you go by quantum physics, this could actually be a possibility. Penrose writes that a

> "... *potential possibility* need only play a role *as a counterfactual*, according to quantum theory, for it to have an *actual* effect."[142]

Now that's interesting: a mere possibility/probability could therefore play a role as a „counterfactual", i.e. as an opposite, in order to trigger an actual physical effect? In other words, the mere probability that a non-local superposition of all conceivable state descriptions of a certain universe *could* give rise to a „collapsed" one, i.e. a

---

[142] Roger Penrose, „Shadows of the Mind. A Search for the missing Science of Consciousness", Vintage Books, London, 1994, p. 383, my emphases except for „actual".

localised one, e.g. in the form of an existentially concretised physical universe, could already be sufficient to actually trigger precisely this effect?

Boom goes the dynamite.

II
Evolution:
Spatiotemporally localised unfolding of a non-locally enfolded "superposition" of existential possibilities?

"The possible interference patterns between the standing waves, which we know as atoms, determine what kind of molecules the atoms can form and thus what kind of chemical systems can emerge. The interference pattern of the molecules in turn determines the possible types of intermolecular interactions, including the complex interactions that form the basis of life."

Ervin Laszlo, "Science and the Akashic Field. An Integral Theory of Everything", Inner Traditions, Vermont, 2007, p. 28

"Evolution is first and foremost a process of actualisation of potential traits under the influence of different contexts, which can exert their influence one after the other."

Diederik Aerts, Massimiliano Sassoli de Bianchi, "Quantum Perspectives on Evolution", p. 9

Excuse me? Dear author, can you perhaps explain this in more detail? What is that supposed to mean: "Spatiotemporal unfolding of a non-local superposition of existential possibilities"? To be honest, I'm not so sure myself. This is not at all surprising, as the areas of research from which I have drawn the information that inspired me to formulate this are still very young in comparison, not to say in the "embryonic stage". In this second, much shorter part of the book, I will try to give a satisfying explanation for this concept. It is much shorter than the first part because the terms and concepts of quantum physics which underly it have already been presented in detail in the first part.

As we saw in the first part of the book, a mechanistic-deterministic view of the world dominated the natural sciences for a very long time: whether in physics, biology, the behavioural sciences and, of course, in evolutionary science. Quantum physics now seems almost like a blink of an eye: it can be traced back - at least roughly - to the beginning of the 20th century. Although it has of course been developed further since then, undergone numerous additions and been enriched by new models, it is still a 'childlike' science. Only recently have fields of research such as quantum biology been added, and evolutionary biology is now also trying to find more precise explanations for the development of life using quantum physical models.

> *"It is the inability to go beyond the [biochemical] mechanistic framework that leads people to persist in asking which parts [of the body] are in control or giving instructions or information. The challenge for all of us is to rethink information processing in the context of the [quantumly] coherent organic whole.'*[143]

Another problem lies in our conventional concept of time. In earlier times, people looked for a kind of original or starting state, an initial cause, which then - in the deterministic view of the world - set in motion a **linear** development of successive individual steps or stages up to a certain final state. However, there are already ideas of evolution - developed with the help of quantum physics models - that question precisely this linearity: Instead of a simple causal chain „cause -> linear development -> effect" (or „final state"), it could be a far more dynamic process of „actualisation"

---

[143] Mae-Wan Ho, Julian Haffegee, Richard Newton, Yu-ming Zhou, John S. Bolton and Stephen Ross, Bioelectrodynamics Laboratory, and Physics Department Open University, U.K., Bioelectrochemistry and Bioenergetics 41, p. 81-91, 1996.

of potential causes, in which - and continuously - the results of the realisation of this „potentia" - let's just call it „diversity of possible blueprints" here - is quasi „matched" with these „blueprints". In other words: It is not a one-sided process that unfolds „into reality", but rather - to put it very simply - there could be a „feedback" between the „potentia", i.e. the wave or probability function of a possible existential concretisation qua state vector reduction („collapse of the wave function") with „itself". In this sense, there would be an exchange of information between the two levels - the local and the non-local, the possible and the realised - whereby it would again have to be asked to what extent such a strict separation - in the sense of both levels being isolated from each other (i.e. as „classical" dualism) - could be maintained in a meaningful way at all.

Penrose arrives at a similar idea, namely in the form of a 'learning system':

> „We must now ask: What is the nature of these learning processes? We imagine that our **learning system is in an external environment in which the way the system acts in this environment is constantly changed by how its environment has reacted to its previous actions.** There are essentially two factors. The external factor is the way this environment behaves and how it reacts to the system's actions. The internal factor is the way in which the system itself changes its own behaviour in response to these changes in its environment."[144]

I will try to answer the question of the evolutionary stage at which such „learning systems" could emerge below. In my opinion, this depends crucially on the degrees of freedom of a system - and thus inevitably on the complexity of the system (e.g. a life form such as humans), but more on this in a moment. A stone certainly does not fall into the category of a „learning system", whereas humans do - even if some might now sarcastically remark that, given the sad and barbaric fact that wars are still being waged even in the early 21st century, it is hardly possible to speak of a learning process. Others might point out that it could be a case of transposed figures, so that we are living in the 12th century instead of the 21st. I sympathise with this idea, but this book is not intended to be comedy.

Since our universe is extremely complex - you only have to think of the breathtaking complexity of organic life forms that arose in it - the deterministic world view simply assumes that the cause, i.e. the initial state, must have been just as complex. However, this is not necessarily true. As you learnt in the first part of this book, even

---

[144] Roger Penrose, „Shadows of the Mind. A Search for the missing Science of Consciousness", Vintage Books, London, 1994, p. 151, my emphases.

a relatively simple initial condition can lead to very complex systems as it develops over time. Since we have already learnt about many quantum physics concepts and terms in the first part of the book, the following quote from a scientific article explaining this context should not cause any further problems of understanding:

> „After this quantum (and quantum-like) digression, let us return to our original point about the possible inadequacies of the Darwinian explanation of natural selection. Our hypothesis is that selection in nature occurs both as selection of **actual** traits and as **selection through the realisation of potential traits**, and that **the former can indeed be understood as a special case of the latter**. More specifically, we can think of the evolution of an entity in general as the result of its interaction with a context. In other words, **evolution is primarily a process of actualisation of potential properties under the influence of different contexts** that can exert their influence in succession [...]
>
> Among them, there are the indeterministic ones, usually called (quantum) measurements in physics (although physicists do not consider them as contexts for historical reasons), which, as we have explained using the example of quantum machines, **do not only** apply **to microscopic entities;** and there are the deterministic ones, usually not called measurements, but simply developments, in which the process of change is usually assumed to be continuous.
>
> In quantum mechanics, the first possibility is described by the projection postulate and the associated Born's rule of **probability assignment**, the second by considering a so-called time-dependent unitary evolution operator that obeys the Schrödinger equation (or Dirac equation) and **acts on the state of the entity** to describe its change over time. Note, however, that a deterministic state change in the final analysis **can also be understood as a measurement**, and a measurement **with only one possible outcome**, so that a **continuous state change can also be described as a recursive application of these particular measurement processes with one outcome**. It is worth noting that this view of change as context-driven actualisation of potential [...], which **describes the evolution of entities as processes of actualisation of potential properties through repeated interactions with**

> *multiple contexts [...], allows the construction of more general models of variation of forms than the standard Darwinian view allows.*"[145]

Okay, that was perhaps a bit abstract again. To make it easier to understand: A „deterministic" change - i.e. one in the world of classical physics, meaning a change in a well-defined state, for example in the form of a concrete organism such as an amoeba, a dragonfly or a sloth - could also be understood as a „measurement" („observation"). A „continuous change of state" - e.g. in the form of an evolutionary development process - could thus be understood as a „recursive", one could also say: iterative application of this „measurement", i.e. realisation process from a state of potentiality or probability (non-locality, quantum-physical superposition, simultaneity of many possible alternative states) to an existentially concretised, unambiguous, clearly defined (in the form of a spatio-temporally localised state).

In other words, assuming this theory is correct, there would not be an original initial state in the sense of classical physics, meaning a state that sets a **linear** unfolding/development process in motion (which would no longer have any direct relationship to its origin), but a ***continuous interaction***, i.e. an interaction between the potential state and its local realisation (whereby „local" would be understood here in the quantum-physical sense, i.e. in contrast to non-local potentiality).

These „updates" - or continuous „updates" of a certain state through repeated interaction with „multiple contexts" - for example a complex, multi-layered „context" called the „natural environment" - could then lead to what we refer to as „evolutionary leaps" or „mutations". A highly complex process such as evolution can certainly not be reduced to this factor alone - there could be many other important influencing factors. For example, there is the theory that cosmic radiation could be a mutagenic factor that led to changes in the genetic material of living organisms. However, this theory suffers from the fact that it is unable to explain how purely random mutations can lead to highly targeted changes, as is regularly found in the development of living organisms - in the sense of functional optimisation within the framework of an entire organism, i.e. a „fine-tuning" of the entire organism. In other words: ***why*** should organisms tend to develop every higher forms of structural organization and optimisation of certain organs? Where does the „need" to develop an extremely sophisticated organ like the human eye – or the brain – stem

---

[145] Diederik Aerts, Massimiliano Sassoli de Bianchi, „Quantum Perspectives on Evolution", 2019, https://www.researchgate.net/publication/323162300_Quantum_Perspectives_on_Evolution

from? To shorten this story a bit: It amounts to age-old question if there is a purpose to evolution, i.e. the discussion about a possible evolutionary teleology.[146]

> „If we assume that all cells are cognitive and, as shown in photosynthesis, have the ability to explore possibilities in quantum space before expending resources to realise them physically, then we can imagine how mechanisms for high-speed evolution [...] can operate with unprecedented efficiency."[147]

As you learnt in the first part, the question of the 'hen and the egg' is by no means settled: what came first, matter or consciousness? Did the universe really emerge from a big bang, or to put it another way: are there really no other possible explanations? So is the big bang theory the only one that makes sense? If not, what other explanations could be found? And how plausible are they? Scientists of the materialist faction will firmly assert that the question makes no sense, because it is clear that there can be no other cause than the material or substantial particles that make up everything. This also applies to the phenomenon of human consciousness and thought: everything has a purely material cause.

> „The chemist and physicist Ilya Prigogine, winner of two Nobel Prizes for Chemistry, wrote: »The statistical probability that organic structures and the precisely coordinated reactions that are typical of living organisms arise by chance is zero.«"[148]

If it weren't for the „small" problem of quantum physics, which has mercilessly slid into the democritical world view and that of the Newtonian, mechanistic explanation of the world: Pardon, but the deeper we look into matter, the more we have to realise that we are dealing with, as Hans-Peter Dürr so excellently put it, „the dross of the spiritual". Or as Heisenberg wrote: a „tendency towards realisation". In other words: We are dealing with probabilities („probability" or „wave functions").

---

[146] Max Dresow, Alan C. Love, „Teleonomy: Revisiting a Proposed Conceptual Replacement for Teleology", https://www.ncbi.nlm.nih.gov/pmc/articles/PMC10191995/
[147] Perry Marshall, https://www.sciencedirect.com/science/article/pii/S007961072300041X#bib3
[148] C. Allan Boyles, "God and Quantum Physics", Wheatmark, 2021, p. 132

But first a few words on the history of the theory of evolution.

The first systematic reflections on this can already be found in the early modern period, which

> "link naturalistic theories of earth history, embryological development, and the modification of organic species in response to external changes in living conditions. These connections develop directly from the critique of the pre-existence theory of procreation [...] which reformulated, in different but related ways, the 'mechanistic' epigenetic embryology previously advocated by Descartes and the theorists of atomism. Similar to these discarded theories [...] the existence of equivalent male and female 'seeds' was postulated, which would then combine to form the embryo during sexual conception. The new versions of the eighteenth century supplemented these seventeenth-century accounts with a new role for dynamic notions of matter and inherent organising forces."[149]

For those unfamiliar with the term epigenetics, this biological discipline is concerned with the factors that can influence the development of a cell, whereby these are „external" („epi-„) factors, not „internal" ones, i.e. those that can be traced back to changes in the genetic material of the DNA itself.

The concept of „evolution" looks back on a long history of different theories not only about the origin of species, but also of the world itself - in this respect, overlaps with cosmogonic ideas can sometimes be identified. This should come as no surprise, as anyone attempting to explain how living organisms could have evolved will sooner or later naturally come up against the question of how the „beginning of the world" could have played a part in this process of creation.

Nowadays, „evolution" is generally understood to simply mean the theory of how organic species develop and change over time. Naturally, this involves very long periods of time and the question of which mechanisms or processes led and lead to the emergence of new species. In the first half of the 19th century, the term was used almost exclusively in the embryological sense, i.e. to describe the development of individual embryos. As far as the aforementioned overlaps between evolutionary biology and theories of the origin of the world are concerned, the English philosopher Herbert Spencer, for example, used the term „evolutionary history" in

---

[149] Phillip Sloan, "Evolutionary thought before Darwin", https://plato.stanford.edu/entries/evolution-before-darwin/

1852 to express his idea that the entire universe showed a tendency to develop from initial „homogeneity" to increasing „heterogeneity". In other words: starting with the simplest forms, life increasingly develops in the sense of a differentiation from simple to more complex forms, i.e. those of a higher organisational level.

The American evolutionary biologist Douglas J. Futuyma, born in 1924, gives the following explanation in his book „Evolution":

> „Biological evolution is the change in the characteristics of groups of organisms over generations [...] it includes everything from minor changes in the proportions of different forms of a gene within a population to the changes that led from the earliest organism to dinosaurs, bees, oak trees and humans."[150]

The Canadian ethnologist and evolutionary biologist John Endler, born in 1947, provides a very similar explanation:

> „Evolution can be defined as any directed net change or any cumulative change in the characteristics of organisms or populations over many generations - in other words, descent with modification [...] It explicitly includes both the emergence and spread of alleles, variants, trait values or trait expressions."[151]

In addition to biological evolution, which refers to the emergence and change of species, there is also molecular evolution - which examines the molecular changes in macromolecules such as DNA and RNA.

Analogous to the concepts of God in the first part of the book, there were also corresponding concepts regarding the origin of species very early on. The Greek philosopher Empedocles (ca. 495-35 BC) and the ideas of the Greek atomists provided a natural philosophical foundation on which later considerations were built. In their reflections and speculations on the origins of life, the pre-Socratic speculations were based on random processes that emerged from "atomic chaos". These atomistic theories essentially formed the basis of evolutionary theories that were developed in the western hemisphere in the 15th century - around 1417. As the process of emergence is random, the entire development of life is seen as an undirected process in which the best-adapted species - those that are best able to adapt to the conditions of their immediate living environment and its changes - are

---

[150] Douglas J. Futuyma, „Evolution", 2005, Sunderland, MA: Sinauer Associates
[151] John Endler, "Natural Selection in the Wild", 1986, Princeton, NJ: Princeton University Press.

favoured, while all others are gradually "weeded out", i.e. subject to natural selection.

> „[Francis] *Crick calculated the probability of the random formation of a single protein (from which the first DNA molecule could then form). In all living species, proteins consist of exactly the same 20 amino acids, which are small molecules. The average protein is a long chain of about 200 amino acids selected from these 20 and strung together in the correct order. According to the laws of combinatorics, there is a 1 in 20 chance, multiplied by 200 times, that a single specific protein will be created by chance. This number, which can be written 20^{200} and is roughly equivalent to 10^{260}, is many times greater than the number of atoms in the observable universe (estimated at 10^{50}). These numbers are inconceivable to the human mind. [...]* **Crick concludes that the organised complexity found at the cellular level »cannot have arisen by pure chance«.**"[152]

In 1835, the term "transformism" emerged in French biology to describe the change of species, which took on a different meaning in the 1860s and was eventually replaced by "evolution". This term is generally associated with selection, i.e. natural selection; however, it goes far beyond that and also extends to the question of the origin of organic life, the precise interactions between environment and organism and also to the philosophical question of a possible goal of evolution, i.e. the question of a possible evolutionary teleology (purposefulness).

The speculative descriptions of the ancient Greek atomists were opposed on several levels by the later main currents of the Platonic, Neoplatonic, Aristotelian and Stoic philosophical traditions. One of the most important texts in this regard is Plato's "Timaeus" - a creation myth that assumes a kind of "intelligent craftsman", a "demiurgos" (demiurge = creator), who both establishes the origin of living beings and specifies the lines of development along which they evolve. I will come back to this idea later. By virtue of his reason, this Demiurge organises matter, conceived according to mathematical principles, into a structured universe. The life forms in this cosmos are designed according to eternal, unchanging basic forms –

---

[152] Jeremy Narby, "The Cosmic Serpent. DNA and the Origins of Knowledge", Weidenfeld & Nicolson, 199, p. 75-76, my emphasis.

"archetypes" - which originate from the Platonic sphere of "pure forms". Plato's philosophy can be understood as an alternative to the theory of chance: Not chance, but an intelligence - an organising or structuring reason - forms the basis of all existence in the universe.

> *"... the behavioural pattern of an ant is immensely complex and subtle. Are we to believe that their marvellously effective control systems are not supported by the principle that gives us our own qualities of understanding?"*[153]

For a long period of time, the writings of the Greek physician Claudius Galenus (129-200 AD) had a formative influence on the early biosciences and the later biosciences that developed from them. The idea that the high degree of order and complex organisation in the anatomy of living beings was proof of a rational design was decisive:

> *"These interpretations of 'teleological design' interacted in complex ways with the Jewish, Christian and Islamic biblical concepts of creation [...] A common meaning of the term 'teleology' frequently encountered in discussions of evolution since Darwin - that of an externally imposed design by an intelligent agency (demiurge, nature, God) on pre-existing matter - has its origins in these ancient discussions and is not precisely identified with the biblical concept of creatio ex nihilo [...]."*[154]

Aristotle, on the other hand, was not convinced that it must necessarily have been an act of creation by a "demiurge", a divine authority: The idea of an "external" purpose or ultimate end (= telos, hence teleology: purposefulness) imposed on the world by a God or supreme reason standing above or outside of his creation was replaced by an "internal" one in the sense that things had an inherent causal agency. According to Aristotle, living beings were the "informing soul" (psyxé, from which our modern "psyche" is derived). He rejected the idea of a historical origin of the world and instead spoke of the "eternity of the world order". The embryological development of humans from "primordial matter" is a gradual process that is guided by this very

---

[153] Roger Penrose, „Shadows of the Mind. A Search for the missing Science of Consciousness", Vintage Books, London, 1994, p. 408.
[154] Phillip Sloan, „Evolutionary Thought before Darwin", https://plato.stanford.edu/entries/evolution-before-darwin/

"psyche". As for the origin of this "psyxé", i.e. the "soul", Aristotle surmised either the male parent or the sun as the source.

In the fifteenth century, Greek and Roman atomism entered the Renaissance with the rediscovery of the poem "On the Nature of Things" ("De rerum natura") by the Roman poet and philosopher Titus Lucretius Carus, commonly known as Lucretius (c. 99-94 BC to 55 or 53 BC). This led to a series of cosmological speculations,

> *"which included a naturalistic account of the origin of species that was integrated into a non-teleological materialist cosmology that stood in marked contrast to the surviving scholastic, Aristotelian and Augustinian-Platonic traditions. Until the 18th century, these speculations and those stemming from other atomistic sources [...] were often in the background of novel early modern reflections on the origin of species and their possible change over time."*[155]

The synthesis of natural philosophy and metaphysical considerations by the French mathematician and philosopher René Descartes in his "Principia philosophiae", which appeared around the middle of the 17th century, created a new dynamic for systematic considerations on the origin of the earth and its life forms. However, his cosmology was problematic in that

> *"... he failed to include the origins of living beings in his naturalistic history of creation through natural law.[...] In the »Principles«, the question of a naturalistic explanation of the origin of life and the origin of the individual species is simply ignored."*[156]

One possible answer to this question could be to see the inorganic forms of „matter", which we commonly refer to as „dead", as an expression of an evolutionary process whose goal is to create structures of increasing complexity with the aim of enabling more degrees of freedom or greater freedom in the interaction mechanisms of the structures and - at a later stage of evolutionary development - of the organic organisms with their environment (external influences) in order to ultimately produce life forms which, due to their higher degree of organisation and increasing degrees of freedom, are able to influence their environment and thus even create changed conditions for their own further development. That was certainly a bit

---

[155] Ibid.
[156] Ibid.

abstract again, so here are a few examples to explain it, spiced up with a little humour:

A stone has no degrees of freedom to interact with its environment. It consists of a fixed structure of atoms that are in a certain relationship to each other. For example, silicon dioxide, SiO_2. One atom of silicon, two atoms of oxygen. The stone lies somewhere in the environment, either on land or in water. It cannot move on its own, but can only be moved by external environmental influences. From an evolutionary point of view, the stone would therefore be a „dead end", or perhaps better a „preliminary stage" (in the sense of a continuous process of further development): There is nothing more to it than a stony existence. Nor can it reproduce. Bummer.

A plant has more interaction possibilities, albeit not active ones: it absorbs photons to produce nutrients through photosynthesis, which it needs to survive, it forms roots to extract water from the soil, which it also urgently needs, and it releases a gas (oxygen) back into the environment. It can also reproduce by producing seeds that are either carried by the wind to its immediate surroundings or simply fall to the ground to grow other plants of its species. So it already has a little more freedom than the stone. However, it cannot move under its own power either: it remains where it took root. So it cannot simply change its location. For example, if someone were to spread a poison that kills it, it would have no chance of escaping - except that its seeds have already been spread more widely to allow offspring of its species to grow in other, non-poisoned places. However, it has no influence on this: it is dependent on external factors to distribute its seeds, it cannot „carry them elsewhere" itself.

A fish, on the other hand, has somewhat greater freedom than a plant. Although its existence essentially consists of searching for food or mating partners for reproduction through moving around, it can - in contrast to plants - evade possible threats to its existence, e.g. from natural predators, i.e. react to corresponding environmental signals/stimuli, mostly (but not only) of a visual nature. If someone were to pour poison into the water, it can relocate, i.e. look for a new habitat - although it remains restricted to water as a habitat. In these activities, however, it is „pre-programmed", as it is not in a position to consciously question the basic conditions of its existence. It is genetically fixed to these behaviours. It doesn't write poems, it won't write treatises in philosophy and the natural sciences, it can't compose a symphony and it can't build skycrapers, etc. etc.

A wildebeest in the steppe or desert is dependent on the waterholes it finds in its natural environment. If it moves too far away from a waterhole, for example through migration, and is unable to reach the next one in time, it will die of thirst. So the same applies to the wildebeest: eat or be eaten; avoid natural predators; search for food; reproduce. It has no way of actively or even proactively changing these factors of its environment. It reacts to stimuli and signals from its environment.

Humans, on the other hand, can, for example, invent a „portable water source" - a bottle made of glass or another material that they produce from substances that they find in their environment - so that they do not run the risk of dying of thirst during long hikes. What's more, it can develop materials that do not exist in this form in nature to make its life easier: In the course of his development, man might come up with the idea of forging a plough out of naturally occurring ores to make it easier to cultivate his fields, initially pulled by animals, later by machines that far surpass an ox or a horse in terms of power. This allows him to drastically reduce the time needed to cultivate the fields, which enables him to devote himself to other things, such as learning other skills. He can actively change his environment, for example by creating artificial water sources (wells, underground pipes, etc.) so that he is no longer dependent on the natural sources in his immediate environment. Man can therefore bring the water to himself, he can collect it in rain barrels, he can even produce water artificially by causing oxygen and hydrogen to react. Unlike fish, he can invent a diving suit with an oxygen tank that allows him to stay underwater for long periods of time; he can even build an artificial habitat such as an underwater research station where he can stay for weeks or months at a time. And so on and so forth. And he can even penetrate into space – an otherwise deadly environment.

It should be clear where I'm going with this: humans **have far more degrees of freedom** than a plant, a fish or a wildebeest. Not least because their anatomical structure makes it much easier for them to learn the necessary skills: the fins of a fish are less suitable for making complex tools or machines. And even if a wildebeest had a flash of inspiration and wanted to invent a water bottle: How would it make this container with rather coarse hooves? Humans, however, have this possibility thanks to their natural „tools" - their hands and fingers. They can even invent machines that can produce things that they could never assemble with their hands: e.g. a microchip for a computer. Such machines could also be defined as technologically refined „extensions" of his natural extremities, his mind, his imagination and creativity. They are capable of all this because of their consciousness and thinking.

Firstly, one could ask whether it was possibly this increasing number of degrees of freedom that strongly favoured or even brought about the emergence of consciousness or conscious thinking. However, one could also ask whether it was perhaps the other way round and - from a pan- or cosmopsychistic position, as I take it - whether these abilities are an expression of a spatio-temporally unfolding, universal evolutionary process that strives for an increasingly differentiated and more complex realisation of its potential for self-consciousness, i.e. becoming more self-aware. This would be the more holistic approach.

As I discovered in the course of my research for this book, I am not the first to develop this idea. It was presented in an article published by news agency Reuters in 2023:

> *"WASHINGTON, Oct 16 (Reuters) - When British naturalist Charles Darwin outlined his theory of evolution in his 1859 book »On the Origin of Species,« proposing that biological species change over time through the acquisition of traits that favour survival and reproduction, it sparked a revolution in scientific thinking.*
>
> *Now, 164 years later, nine scientists and philosophers on Monday proposed a new law of nature that **incorporates** the biological evolution described by Darwin as a **living example of a much broader phenomenon that occurs at the level of atoms, minerals, planetary atmospheres, planets, stars and more**. It states that **complex natural systems evolve to states of greater structure, diversity and complexity**.*
>
> *»We see evolution as a **universal process** that applies to numerous systems, **both living and non-living**, that increase in diversity and structure over time,« said mineralogist and astrobiologist Robert Hazen of the Carnegie Institution for Science, co-author of the scientific paper describing the law in the journal Proceedings of the National Academy of Sciences. The »**law of increasing functional information**« states that evolving systems, **biological and non-biological**, always arise from numerous interacting building blocks, such as atoms or cells, and that there are processes - such as cell mutation - that create many different configurations. Evolution occurs when these different configurations are subjected to selection for useful functions.*
>
> *»We have well-documented laws that describe such everyday phenomena as forces, motion, gravity, electricity, magnetism and energy,« says Hazen. »But these laws do not individually or collectively describe or explain why the universe is becoming more diverse and complex at the level of atoms, molecules, minerals and more.« In stars, for example, only two*

elements - hydrogen and helium - were the main components of the first generation of stars after the Big Bang around 13.8 billion years ago, which created the universe.

This first generation of stars forged about 20 heavier elements such as carbon, nitrogen and oxygen in the thermonuclear fusion cauldrons of their cores, which were ejected into space when they exploded at the end of their life cycle. The next generation of stars, formed from the remnants of the previous generation, forged almost 100 more elements in a similar way. On Earth, living organisms became increasingly complex, culminating in the pivotal moment when multicellular life emerged. The authors proposed three universal concepts of selection: the fundamental ability to persist, the durability of active processes that can enable evolution, and the emergence of novel traits as an adaptation to an environment. [...] »The importance of formulating such a law is that it provides a new perspective on why the various systems that make up the cosmos evolve the way they do [...],« added co-author Jonathan Lunine, chair of the Department of Astronomy at Cornell University [...]."[157]

The development from simple elements to higher-level elements and the combination of these elements to form more complex structures such as crystals crystals, minerals, ores, gases, liquids such as water - which is known to be of fundamental importance for the development of organic life - and even more complex molecules up to organic, i.e. carbon-based compounds such as carbohydrates, enzymes and proteins and later a highly developed molecule such as deoxyribonucleic acid, commonly known by its abbreviation DNA, for the transmission of genetic information (blueprints) for „archiving" and passing on could therefore be an expression of a self-organising process.

„*In fact, there are no detailed Darwinian explanations for the evolution of any basic biochemical or cellular system, only a variety of wishful speculations. It is remarkable that Darwinism is accepted as a satisfactory explanation for such a vast subject - evolution -* **without a thorough examination**

---

[157] Will Dunham, "Scientists propose sweeping new law of nature, expanding on evolution", October 16th, 2023, https://www.reuters.com/science/scientists-propose-sweeping-new-law-nature-expanding-evolution-2023-10-16/, my emphases.

of how well its basic theses work in elucidating specific cases of biological adaptation or diversity"[158]

It is hardly surprising that quantum physics was used to clarify the question of how „natural selection" actually works, i.e. evolution as a process of the development of organisms - a question that Darwin's theory does indeed not really answer. However, one should not be too hard on Darwin, as he developed his theory at a time when quantum physics and related or derived disciplines such as quantum biology did not yet exist.

The most exciting of these questions is: How do cells or cell clusters actually „know" *how* they have to organise themselves in order to form higher and highly functional organisms? A simple example: How do cells know that in water it naturally makes much more sense to form fins instead of hands with fingers, as they can displace more water per unit of time and thus enable much better, more efficient locomotion? This question should not be taken too lightly. The most obvious answer is: yes well, that's perfectly logical! Evolution simply „knows" that this makes more sense and „acts" accordingly.

> *"Numerous sources show that all living cells are cognitive. [...] Causation is not exclusively top-down nor bottom-up, but a continuous set of feedbacks which together make the »organized whole« which expressly characterizes the seamless, intentional macroorganism, i.e. there is no privileged level of causation."*[159]

All well and good, but **where from**? What exactly is the information that is exchanged between water and cells, for example, in order to set this development process in motion? After all, the most widespread explanation even today is that it is all a purely randomised process. In this case, however, if this process had proceeded exclusively according to the principle of „trial and error", there would have to be a lot more corresponding intermediate stages that indicate that organisms

---

[158] James Shapiro, Microbiologist, 1996, National Review, 16. September 1996, p.64, my emphasis.

[159] Perry Marshall, https://www.sciencedirect.com/science/article/pii/S007961072300041X

that were absolutely unable to survive in water were actually „tried out" in that medium first.

Since quantum theory, as already mentioned, is the best to date when it comes to describing matter at the atomic and subatomic level, there have of course already been corresponding research and explanatory approaches. Not only quantum biology is developing such approaches, but also biophysics:

> "Darwin's theory of natural selection hinges on genetic variations, survival, and reproduction. It favours organisms with advantageous traits, thereby propelling evolution. Conversely, the principle of quantum superposition illuminates the peculiar ability of quantum entities, such as electrons or molecules, to exist in multiple states simultaneously – a concept that defies classical physics. [...] Our principle opens up promising avenues for exploration. So, what other predictions might arise from this marriage of evolution and quantum physics? There are several:
>
> - **Quantum-Mechanical Communication:** Living systems, such as a swarm of single-celled bacteria, **may harness quantum communication for efficient and covert signalling**, increasing their survivability;
>
> - Quantum-Enhanced Sensing: Organisms could leverage quantum phenomena to develop hyper-sensitive sensors **for detecting subtle environmental changes;**
>
> - Quantum Resilience: Under the guiding hand of natural selection, some life forms may have evolved mechanisms to withstand quantum decoherence (the collapse of quantum properties when particles interact with their surroundings), ensuring the persistence of quantum advantages in the face of adversity;
>
> - Quantum Bio-Order Parameters: **The shapes of biomolecules can create emergent quantum fields that produce higher levels of cellular corporation.** A swarm of bacteria (a collective), for example, creates many clones of itself with mutations – this can only happen if there is a collective. If an environmental pressure destroys all but one of the bacteria with the traits for survival then that one survivor

will clone itself into a new collective."[160]

In other words, information, its exchange and „processing", is a very important part of the physical world. Which, as you know from the first part of this book, is not just information in the conventional or „vernacular" sense of the word, meaning a passive form of it that can be just recalled, but as an active one – as a dynamic process of „in-formation":

> *„The idea that information is present throughout nature is a recurring theme in cultural history, but it is new to Western science. It requires the recognition that information is not an abstract concept: as »in-formation« it has a reality of its own. It is part of the physical universe. And because it occurs everywhere in nature, it is best understood as an extended field."*[161]

And:

> *„The experiences of Apollo astronaut Edgar Mitchell during his stay in space led him to a similar conclusion. According to Mitchell, information is part of a »dyad«, the other part of which is energy. It is part of the very substance of the universe. Information is everywhere, from the beginning. The quantum vacuum is the »holographic information mechanism that records the historical experience of matter«. The information meant here is obviously active: it is »in-formation«. The only question is how this mechanism of information functions in the quantum vacuum: How does it record the »historical experience of matter«?"*[162]

It is therefore not surprising that information theory has also been used to develop approaches for a better explanation of evolutionary processes:

> *„Despite the obvious advantage of simple life forms capable*

---

[160] Prof Stephen Alexander, "This bold new theory of »quantum weirdness« could rewrite the story of evolution", October 28th, 2023, https://www.sciencefocus.com/comment/quantum-weirdness-force-of-life, my emphases.

[161] Ervin Laszlo, "Science and the Akashic Field. An Integral Theory of Everything", Inner Traditions, Vermont, 2007, p. 73.

[162] Ibid., p. 67

of rapid replication, living systems have reached different levels of cognitive complexity in terms of their potential to cope with environmental uncertainty. Given the unavoidable costs associated with recognising and adaptively responding to environmental cues, we hypothesise that the potential to predict the environment may overcome the costs associated with maintaining costly, complex structures. [...] The idea of autonomy and the fact that predicting the future requires some kind of computation suggest that a coherent theory of the complexity of life must involve reproducing individuals (and eventually populations) **and information** *(they must be able to predict future environmental states)."*[163]

In recent years, there has been an increasing amount of research into the question of whether and, if so, what connection there is between „in-formation", i.e. a dynamic „information processing" and the development of the universe through to organic life forms, and how it could be described on the quantum level. In particular, the question of whether an exchange could take place between the local and non-local level is being analysed.

"Several independently developed approaches [...] to a wave theory of genes have been developed. Here, recent novel experiments, carried out in Moscow, at the Institute of Control Sciences of the Russian Academy of Sciences, are reported in confirmation of this theory. The theory changes the accepted notion about the genetic code essentially, asserting :

- that the evolution of biosystems has created genetic »texts«, similar to context dependent texts in human languages, shaping the text of these speech-like patterns,

- that the chromosome apparatus **acts simultaneously both as a source and receiver of these genetic texts**, *respectively decoding and encoding them and*

- that the chromosome continuum of multicellular organisms is analogous to a static-dynamical multiplex **time-space holographic grating, which comprises the space-time of an**

---

[163] Luís F. Seoane, Ricard V. Solé, "Information theory, predictability and the emergence of complex life", https://royalsocietypublishing.org/doi/10.1098/rsos.172221, my emphasis.

organism in a convoluted form. "[164]

Well, that's interesting. I would even say extremely fascinating: the „chromosome continuum" as a kind of holographic grid or „image" that represents the spatiotemporally unfolded organism in a „convoluted" or „enfolded" form. Perhaps this concept could also be imagined as an „idea" or „mind-idea" of the organism in an „enfolded" form, as Walter Russell would have put it? This reminds me of one of the quotes I prefaced this book with, from Manly P. Hall: „... as the seed is the folded tree, so the world is the unfolded god."[165]

At this point, we could once again ask why there might be something about our existence that is „non-computable", as Sir Roger Penrose put it. In his book „Shadows of the Mind", which I discussed in more detail in the first part, Penrose asked about the nature of this level of „non-computability" and used Turing's halting machine and Gödel's incompleteness theorem to explain the problem on the mathematical level. Perry Marshall provided a possible, interesting and plausible explanation in his article „The role of quantum mechanics in cognition-based evolution":

> „*The classical world is governed by laws, which are **deductive**; the quantum world is governed by decisions, which are **inductive**. When the two are combined, they form the main feedback loop of perception and action for all of biology. [...] But by definition, **no system that can do induction can be reduced to maths**. This is why we are unable to explain the origin of the genetic code, model the human mind, accurately predict anything in the soft sciences, or accurately model the evolutionary process. This is not simply because these things are too complex, but because **they are not computational processes. Therefore, a complete mathematical model is not possible**. This could explain why, as Shapiro says, »we lack a detailed account of every major evolutionary transition. The origin of life is the most obvious example, but we have similarly sketchy accounts of how the first cells evolved, how cells developed complex structures like nuclei and mitochondria, how multicellular organisms evolved, and how complex organs like the brain and immune system*

---

[164] Peter P. Gariaev, Uwe Kaempf, Peter J. Marcer, Georg G. Tertishny, Boris Birshtein, Alexander Iarochenko, Katherine A. Leonova, "The DNA-wave Biocomputer", https://userpage.fu-berlin.de/~gerbrehm/nw/wavecomputer.pdf, my emphasis.

[165] Manly P. Hall, „The Secret Teachings of All Ages", H.S. Crocker Company, 1928, p. 169

evolved."[166]

In one of the quotes above (from the paper „The DNA-wave Biocomputer"), a „wave theory" of genes was mentioned. What is it all about? I'm sure you've already guessed: Of course, it also involves quantum physics, namely the well-known wave-particle duality. According to this still very young theory, not only particles, but also more complex biomolecules such as DNA or RNA, proteins etc. could have a quantum non-local wave-character, i.e. be connected to each other and form a systemic whole.

In her paper „Quantum Nonlocal Bonds between living Organisms and their role in Evolution", published in the „Canadian Journal of Pure and Applied Sciences", Firyuza Salikhovna Yanchilina from the Russian Academy of Sciences in Moscow explains this model. She refers to a hypothesis developed by the Russian physicist Vasily Yanchilin, according to which the elementary particles that make up a living organism could be in a very complex quantum state.

> „According to this hypothesis, at the earliest stage of the emergence of complex organic compounds, all of them were in a united quantum state, that is, **in nonlocal interaction between themselves.** This state with quantum entanglement, we will call the quantum Biomass or simply Biomass. Due to a united quantum state, the processes that took place in the complex molecular formations that formed within Biomass were determined by the entire Biomass. Biomass, as a single formation, influenced its own parts thanks to nonlocal bonds and thus promoted the growth of complex molecular formations within itself. The influence of a complex quantum state of Biomass on its parts led to the growing and complicating of these parts. This, in turn, led to the complication of the entire Biomass. As a result of this positive feedback within the Biomass, complex molecular formations had been formed.
>
> The complex quantum state of molecular structures reflected almost all of their evolution, which lasted many hundreds of millions of years. Atoms and molecules that made up these structures were in a single quantum state with all other atoms and molecules of the Biomass.

---

[166] Perry Marshall, https://www.sciencedirect.com/science/article/pii/S007961072300041X, my emphases.

> *As Biomass became more complex, it became more and more separated from the rest of the world. This led to the fact that molecular formations within it also became more complex and they were increasingly separating from each other. Their development to some extent repeated the development of the entire Biomass. However, Biomass continued to be a single whole at a quantum level. Finally, the following thing had happened. Inside itself, Biomass was divided spatially into many parts – unicellular formations, ancestors of modern bacteria.* **But, despite such a spatial division, all parts of Biomass continued to be a single whole at the quantum level.** "[167]

It hardly needs to be mentioned that in the sciences - which is their task anyway - a lively debate is raging about the question of whether evolution is purposeful, i.e. not just chance-based. The emergence of more complex life forms - from the amoeba to man - has not allowed the dispute between the "chance theorists" and the "teleologists" or "teleo-evolutionists" to come to rest. Is there perhaps an "intelligent design" behind life after all, or is this just unscientific rubbish?

Of course, there are flawed arguments on both sides of this debate, which are often based on the fact that certain views that have developed over long periods of time are simply assumed to be firmly established or unquestionable. David Hanke, for example, argued in his essay „Teleology: the explanation that bedevils biology":

> *"Biology is sick. Fundamentally unscientific modes of thought are increasingly accepted, and dominate the way the subject is explained to the next generation. The heart of the problem is that we persist in making (literal) sense of the world **that we now know to be senseless** by attributing subjective values to the objects in it, values that have no basis in reality."*[168]

Amazing, isn't it? Hanke simply claims - which is totally unscientific – „that we know today" that the world has no meaning. First of all, who is „we"? Secondly, where does he get this „knowledge" from? To be honest, such statements frighten me to the core. A single human being just declares existence to be senseless - based on what grounds? Is there any evidence for such a claim? Of course not. It is presumptuous and very arrogant to declare one's own subjective horizon of

---

[167] Firyuza Salikhovna Yanchilina, Russian Academy of Sciences, Moscow, Canadian Journal of Pure and Applied Sciences Vol. 12, No. 3, pp. 4651-4659, October 2018, my emphases.
[168] Brian Miller, "The Return of Teleology to Biology", https://evolutionnews.org/2021/08/the-return-of-teleology-to-biology/, my emphasis.

knowledge as the standard. On top of that, Hanke seems not to have noticed the blatant contradiction, which of course consists of accusing others of approaching the world with „subjective values", while he himself does exactly the same thing: confusing his subjective level of knowledge with an objective reality, i.e. generalizing a preliminary, by no means definitive state of knowledge.

What's more, it's not just a scientific question, but also a philosophical one: if the world is supposed to be completely meaningless or make no sense, why live at all? What kind of world view is it that tells people that the world doesn't make the slightest bit of sense? If everything's meaningless, why not just commit collective suicide? Needless to say that such a worldview is senseless in itself.

Furthermore, it seems to me that there is no real problem here in the sense (pun not intended) that it could be one of „objective reality". It is merely one of the imprecision of human conceptual thinking. It is a pure linguistic problem: what exactly does the word „sense" mean? This is why I emphasised the immense importance of the precision of concepts at the beginning of the book. If this question seems silly to you, perhaps the following logical (and slightly humorous) consideration will help:

What sense would an existence have that allows living beings to completely deny it? This is of course a contradiction, because this fact alone - that living beings capable of thinking can make the assertion of senselesness - could already be defined as sense: Living beings exists to be able to make the claim that the world is senseless. In principle, such „problems" fall into the famous category of the Cretan who says: „All Cretans are liars".

To put it more precisely and in my humble opinion: The fact of existence alone already proves at least one sense - namely that of existing. That is trivial. If there were no sense in existence, nothing would exist.

Final remarks

There is sometimes a false impression that the emergence of new scientific theories disproves older ones or even renders them superfluous. This may be true in individual cases, but they are extremely rare. Instead, the natural sciences in particular must be seen as an evolutionary process of acquiring knowledge, in which older models are not rendered obsolete, but merely expanded and supplemented.

Newton's "classical" mechanistic view of the world was by no means rendered completely superfluous by Einstein's general and special theories of relativity and the emergence of quantum physics. Many of Newton's concepts still have practical applications today. What did change, however, was the understanding of what we commonly call "reality". Instead of a static space in which celestial bodies orbit, Einstein introduced the concept of a spacetime whose shape is influenced by the presence of masses: very large masses are able to bend spacetime so much that even light flies around them in curves instead of straight lines.

Furthermore, energy can be transformed into mass and vice versa, mass can become energy. In other words, and to cut a long story short: previously fixed boundaries between mass, energy, time and space, which were assumed to be immovable, became more "permeable", they became more flexible and "blurred" into one another.

Quantum physics once again shattered such boundaries: instead of previously "solid" particles, the physicists who were significantly involved in the development of this theory - such as Heisenberg and Bohr - regularly struggled with their new findings, as they shook their classical physical world view: the so-called solidity was rather a question of frequence, of vibratory rates and "probabilities", a "tendency to exist". Here is an example from everyday experience: You will certainly be familiar with the phenomenon of the supposedly "stationary spokes" which can be observed in movies. If the wheel of a bicycle turns at a certain speed, the spokes appear to "stand still". This is of course not the case; it is merely an optical illusion resulting from the image refresh rate of the movie, the rotational speed of the wheel and the ability of our eyes to perceive image sequences above a certain speed no longer as individual images but as a "stream of images". Analogous to this - as you have learnt from this book - there is the concept of matter as a "standing wave" - but this is just a question of the frequency with which "matter" vibrates.

Heisenberg wrote in his book "Quantum Theory and Philosophy" that he and Bohr sometimes spent nights racking their brains over the new models of quantum physics and regularly "almost despaired". Of course, acceptance was also difficult because - as quantum physicists like to say - the so-called "quantum weirdness" has a "counterintuitive" character: it contradicts our everyday experience.

In my opinion, however, it will only be a matter of time before this "weirdness" disappears and is recognised as completely "normal". Why? Because our thinking today is still too much characterised by the old scientific models and paradigms.

Many readers will be familiar with Plato's „Allegory of the Cave'". Plato, who we encountered during a brief discussion of the question of why humans are actually capable of mathematical understanding (see the quote from Penrose's book „Shadows of the Mind"), posed a very similar question. According to him, there is a sphere or plane of existence of „pure forms", whereas the physical forms derived from them - those of our reality - are merely shadows of these perfect forms. The people in this cave are chained in such a way that they cannot see the perfect forms behind them. They only see their shadows, which appear imperfect, „frayed", somehow „incomplete" - caused by the flickering of a fire that casts light on the cave wall.

Allow me to conclude this book with another daring interpretation of this parable using quantum physical terms and based on everything that has been compiled in this book:

The cave is the „material", physical world. The chains would metaphorically stand for the fact that we are forced to think in dimensional terms, that all our perception and thinking is „imbued" with dimensionality - thus characterised by the physical world of „shadows".

The „perfect" or „pure" forms are the possibilities/probabilities of realised, i.e. existentially concretised forms. They exist in the non-local and in the form of probability functions that have not yet been localised, i.e. have not yet been physically „fixed", i.e. realised, by the collapse of their probability functions. But what could the „flickering of the fire" stand for? For the oscillations or vibration rates that underlie „matter" and thus these existentially concretised forms.

Only in the non-local, only as pure probability are the forms „perfect" - insofar as one understands this word as „not yet realised", as mere imagination or „spirit-idea" in Russell's sense. And in the imagination - as we all know - everything is of course always perfect. A composer has the „perfect" symphony in his head - until he sits down at his piano and has to realise his „spiritual idea" of a composition in the form of a score. In the „friction" with the real, physical world - in the local - a series of tones unfolds that previously existed only as a probability in his head - as the „superposition" of a symphony, as a collection of vague possibilities.

You know what? I'd like to finish off by triggering those a little bit who immediately get hair loss, severe skin irritations and diarrhoea from words like "God". What if the "omniscience" and "omnipresence of God" were to be understood as the probability function in the non-local? Would that be so surprising? After all, the probability function describes **all** conceivable/possible states in superposition, i.e. simultaneously and everywhere. Before a localisation took place, i.e. a "collapse" into an existential concretisation called "world", it would be no wonder that "God" - understood as an all-encompassing possibility - would be everywhere and "omniscient", because "he" can realise himself at any conceivable time and in any conceivable place by means of a self-observation or "measurement".

Take it easy. Take a deep breath. I was only speculating.

I don't know whether there is a „God" - at least in the form in which he is presented by the religions of this world. But what I do **think** I know, what I feel with every cell in my body and what I personally am absolutely convinced of: That there is a „supreme mind", a „spirit", a consciousness that from a „superposition" of all conceivable possibilities to create a universe - a „superposition state" of all possible states in simultaneity - could only bring itself into conscious experience through realisation - through existential concretisation - and seeks to „rediscover" itself through an evolutionary process of **becoming** self-aware in its local „shadow".

To put it more poetically: If one imagines the emergence of the cosmos as a kind of „centrifugal force" that hurls „spirit-idea" out of itself in the form of a physical universe, then the evolving human spirit could be part of a „centripetal force" searching for its origin, or, if you will, trying to „find its way home". In this sense, the entire universe could be a posed question:

„What am I and why?"

> „That is why man is man's greatest wealth, because what he most urgently needs is another person to whom he can give of his own overflowing self in order to become richer himself through what flows back to him."[169]

---

[169] Walter Russell, „Eine neue Vorstellung vom Universum", Genius-Verlag 2019, p. 163

Index

"Bootstrap" theory 8, 89
Ain 20, 23, 24, 30, 67, 68
Ainsoph 23
Ainsophaur 23
Albert Einstein 5, 9, 46
Alexander Iarochenko 122
Alfred North Whitehead 90
Anaxagoras 27
Ashvaghoosha 10
Avatamsaka Sutra 31
Baruch de Spinoza 29
Big Bang theory 5, 13, 14, 15, 18, 22, 64
Boris Birshtein 122
Brahman 30, 31, 87
Brian Miller 124
Bruce D. Curtis 91
C. Allan Boyles 16, 19, 32, 62
Christof Koch 12, 73
Chuang Tzu 19
Cosmopsychism 48
Daniel Schenz 71
David Bohm 47, 58, 65, 80, 88
Diederik Aerts 102, 107
Douglas J. Futuyma 110
Eddy Keming Chen 85
Edward N. Zalta 12, 72, 74, 75
Eheieh 23
Ervin Laszlo 4, 15, 16, 18, 27, 48, 49, 60, 64, 80, 81, 84, 120
Euclidean geometry 9
Firyuza Salikhovna Yanchilina 123, 124
Francesco Patrizi da Cherso 27
Friedrich Wilhelm Joseph von Schelling 32
Fritjof Capra 6, 7, 8, 13, 30, 31, 34, 36, 48, 50, 51, 64, 66, 81, 83, 89
Galileo Galilei 51
Geoffrey F. Chew 9
Georg G. Tertishny 122
Georg Wilhelm Friedrich Hegel 34
Giulio Tononi 12, 73
Gottfried Wilhelm von Leibniz 31
Hans-Peter Dürr 32, 68

Harald Atmanspacher 12, 72, 74, 75
Henri Bergson 33
Heraclitus 49, 50
Herbert Spencer 33
Hermes Trismegistos 36
Hinduism 17, 30, 36, 77, 79, 82
Hiram Abiff 40
Ilya Prigogine 108
J.A. Wheeler 7
J.E. Cirlot 40
J.J. Hurtak 91
James Hartle 84
James Shapiro 118
Jeremy Narby 40, 111
Johann Gottlieb Fichte 32
Johannes Kepler 82
John Archibald Wheeler 7, 58
John Endler 110
John Horgan 59
John S. Bolton 104
Joseph Selbie 58, 65, 99
Julian Haffegee 104
Kabbalah 16, 20, 24, 31, 87, 90, 93, 100
Karl Tate 15
Katherine A. Leonova 122
logos 50
Luís F. Seoane 74, 121
Mae-Wan Ho 104
Manly P. Hall 3, 11, 20, 23, 24, 25, 27, 29, 30, 33, 35, 39, 40, 41, 122
Maria Popova 1
Massimiliano Sassoli de Bianchi 102, 107
Max Born 10, 83, 84
Max Planck 32, 98
Metaverse 15, 16, 17, 18
Michio Kaku 85, 86
Nagarjuna 42
Niels Bohr 35, 58
Nonlocality 19, 32, 39, 40, 52, 64, 65, 66, 67, 68, 69, 80, 88, 92, 95, 99, 100
Panpsychism 25, 26, 40, 43, 44, 49, 72, 75

Parmenides 49, 50
Paul Dirac 7, 55
Perry Marshall 4, 108, 118, 122, 123
Peter J. Marcer 122
Peter P. Gariaev 122
Philip Goff 25, 26, 43, 45
Phillip Sloan 109, 112
Plato 27, 35, 56, 60, 69, 88, 127
Poimandres 37, 38
Pythagoras 35
Quantum Entanglement 64, 70, 93
Quantum Field 7, 9, 77, 81
Quantum Vacuum Energy 64, 90
Rhawn Gabriel Joseph 69, 76, 77
Ricard V. Solé 74, 121
Richard Newton 104
Richard P. Feynman 3, 70
Robert Van Gulick 72
Roger Penrose 52, 55, 57, 60, 61, 69, 92, 99, 100, 105, 112, 122

Sean Allen-Hermanson 25, 26, 43, 45
Sheldon Goldstein 43
S-matrix 7, 8, 89
Stephen Alexander 120
Stephen Hawking 1
Stephen Ross 104
Swami Vivekananda 16
Taoism 31, 77, 79
The Secret Teachings of All Ages 20, 23, 27, 33, 35, 37, 39, 41
Uwe Kaempf 122
Walter Russell 22, 28, 29, 47, 64, 75, 76, 78, 79, 87, 88, 94, 128
Werner Heisenberg 7, 35, 56, 63, 77, 78
Will Dunham 117
William Seager 25, 26, 43, 45
Xenophanes 27
Yu-ming Zhou 104

www.ingramcontent.com/pod-product-compliance
Lightning Source LLC
Chambersburg PA
CBHW071831210526
45479CB00001B/90